Discrete Mathematics and its Applications

Discrete Mathematics and its Applications

Joshua Walters

MURPHY & MOORE
www.murphy-moorepublishing.com

Published by Murphy & Moore Publishing
1 Rockefeller Plaza,
New York City, NY 10020, USA
www.murphy-moorepublishing.com

Discrete Mathematics and its Applications
Joshua Walters

International Standard Book Number: 978-1-63987-166-7 (Hardback)

Cataloging-in-Publication Data

Discrete mathematics and its applications / Joshua Walters.
p. cm.
Includes bibliographical references and index.
ISBN 978-1-63987-166-7
1. Discrete mathematics. 2. Mathematics. 3. Computer science--Mathematics. I. Walters, Joshua.
QA297.4 .D57 2022
518.25--dc23

Contents

Preface

The study of mathematical structures which are discrete in nature rather than being continuous is termed as discrete mathematics. The fundamental objects of this discipline such as integers, graphs and logic statements do not vary smoothly but have distinct and separated values. It also deals with countable sets. The subject of discrete mathematics can be further classified into information theory, set theory, combinatorics, probability, graph theory, number theory, game theory, geometry, topology and operations research. It finds extensive applications in the fields of programming language, cryptography and software development. This book presents the complex subject of discrete mathematics in the most comprehensible and easy to understand language. The topics included herein on discrete mathematics are of utmost significance and bound to provide incredible insights to readers. The book is appropriate for those seeking detailed information in this area.

To facilitate a deeper understanding of the contents of this book a short introduction of every chapter is written below:

Chapter 1- The study of structures in mathematics which are basically discrete instead of continuous is called discrete mathematics. The integers and graphs studied within this field have distinct and separated values. This chapter has been carefully written to provide an easy introduction to the the varied facets of discrete mathematics.

Chapter 2- The branch of mathematics which focuses on the networks of points connected by lines is known as graph theory. The different types of graphs studied within this field are null graph, simple graph, trivial graph and undirected graph. All these diverse types of graphs have been carefully analyzed in this chapter.

Chapter 3- A set refers to an unordered collection of objects. These objects are called members of the set. A few types of sets are finite and countably infinite sets. A function is used to assign one member of a related set to each member of a particular set. The topics elaborated in this chapter will help in gaining a better perspective about sets, relations and functions.

Chapter 4- The branch of mathematics which deals with integers and integer value functions is called number theory. A few of its major branches are analytic number theory and algebraic number theory. This chapter closely examines these facets of number theory to provide an extensive understanding of the subject.

Chapter 5- The area of study within mathematics which deals with counting as well as some properties of finite structures is called combinatorics. Generating function is a method of encoding an infinite series of numbers by treating them as coefficients of a formal power series. This chapter discusses in detail the different aspects of combinatorics and generating function such as pigeonhole principle and inclusion-exclusion principle.

I would like to share the credit of this book with my editorial team who worked tirelessly on this book. I owe the completion of this book to the never-ending support of my family, who supported me throughout the project.

Joshua Walters

1

Discrete Mathematics: An Introduction

The study of structures in mathematics which are basically discrete instead of continuous is called discrete mathematics. The integers and graphs studied within this field have distinct and separated values. This chapter has been carefully written to provide an easy introduction to the the varied facets of discrete mathematics.

Discrete mathematics is mathematics that deals with discrete objects. Discrete objects are those which are separated from (not connected to/distinct from) each other. Integers (aka whole numbers), rational numbers (ones that can be expressed as the quotient of two integers), automobiles, houses, people etc. are all discrete objects. On the other hand real numbers which include irrational as well as rational numbers are not discrete. Between any two different real numbers there is another real number different from either of them. So they are packed without any gaps and cannot be separated from their immediate neighbors. In that sense they are not discrete.

Let us first see why we want to be interested in the formal/theoretical approaches. Some of the major reasons that we adopt formal approaches are 1) we can handle infinity or large quantity and indefiniteness with them, and 2) results from formal approaches are reusable. As an example, let us consider a simple problem of investment. Suppose that we invest $1,000 every year with expected return of 10% a year. How much are we going to have after 3 years, 5 years, or 10 years? The most naive way to find that out would be the brute force calculation.

Let us see what happens to $1,000 invested at the beginning of each year for three years. First let us consider the $1,000 invested at the beginning of the first year. After one year it produces a return of $100. Thus at the beginning of the second year, $1,100, which is equal to $1,000 * (1 + 0.1), is invested. This $1,100 produces $110 at the end of the second year. Thus at the beginning of the third year we have $1,210, which is equal to $1,000 * (1 + 0.1)(1 + 0.1), or $1,000 * $(1 + 0.1)^2$. After the third year this gives us $1,000 * $(1 + 0.1)^3$.

Similarly we can see that the $1,000 invested at the beginning of the second year produces $1,000 * $(1 + 0.1)^2$ at the end of the third year, and the $1,000 invested at the beginning of the third year becomes $1,000 * (1 + 0.1). Thus the total principal and return after three years is $1,000 * (1 + 0.1) + $1,000 * $(1 + 0.1)^2$ + $1,000 * $(1 + 0.1)^3$, which is equal to $3,641.

One can similarly calculate the principal and return for 5 years and for 10 years. It is, however, a long tedious calculation even with calculators. Further, what if you want to know the principal and return for some different returns than 10%, or different periods of time such as 15 years? You would have to do all these calculations all over again.

We can avoid these tedious calculations considerably by noting the similarities in these problems and solving them in a more general way. Since all these problems ask for the result of investing a certain amount every year for certain number of years with a certain expected annual return, we use variables, says A, R and n, to represent the principal newly invested every year, the return ratio, and the number of years invested, respectively. With these symbols, the principal and return after n years, denoted by S, can be expressed as,

$$S=A(1+R)+A(1+R)^2+...+A(1+R)^n.$$

As well known, this S can be put into a more compact form by first computing S - (1 + R)S as,

$$S=A\left((1+R)^{n+1}-(1+R)\right)/R.$$

Once we have it in this compact form, it is fairly easy to compute S for different values of A, R and n, though one still has to compute $(1 + R)^{n+1}$. This simple formula represents infinitely many cases involving all different values of A, R and n. The derivation of this formula, however, involves another problem. When computing the compact form for S, S - (1 + R)S was computed using $S = A(1 + R) + A(1 + R)^2 + ... + A(1 + R)^n$. While this argument seems rigorous enough, in fact practically it is a good enough argument, when one wishes to be very rigorous, the ellipsis ... in the sum for S is not considered precise. You are expected to interpret it in a certain specific way. But it can be interpreted in a number of different ways. In fact it can mean anything. Thus if one wants to be rigorous, and absolutely sure about the correctness of the formula, one needs some other way of verifying it than using the ellipsis. Since one needs to verify it for infinitely many cases (infinitely many values of A, R and n), some kind of formal approach, abstracted away from actual numbers, is required.

Suppose now that somehow we have formally verified the formula successfully and we are absolutely sure that it is correct. It is a good idea to write a computer program to compute that S, especially with $(1 + R)^{n+1}$ to be computed. Suppose again that we have written a program to compute S. How can we know that the program is correct? As we know, there are infinitely many possible input values (that is, values of A, R and n). Obviously we cannot test it for infinitely many cases. Thus we must take some formal approach.

Related to the problem of correctness of computer programs, there is the well-known "Halting Problem". This problem, if put into the context of program correctness, asks whether or not a given computer program stops on a given input after a finite amount of time. This problem is known to be unsolvable by computers. That is, no one can write a computer program to answer that question. It is known to be unsolvable. But, how can we tell it is unsolvable? How can we tell that such a program cannot be written? You cannot try all possible solution methods and see they all fail. You cannot think of all (candidate) methods to solve the Halting Problem. Thus you need some kind of formal approaches here to avoid dealing with an extremely large number (if not infinite) of possibilities.

Discrete mathematics is the foundation for the formal approaches. It discusses languages used in mathematical reasoning, basic concepts, and their properties and relationships among them. Discrete mathematics is also concerned with techniques to solve certain types of problems such as how to count or enumerate quantities.

Since the time of Isaac Newton and until quite recently, almost the entire emphasis of applied mathematics has been on continuously varying processes, modeled by the mathematical continuum and using methods derived from the differential and integral calculus. In contrast, discrete mathematics concerns itself mainly with finite collections of discrete objects. With the growth of digital devices, especially computers, discrete mathematics has become more and more important.

Discrete structures can be counted, arranged, placed into sets, and put into ratios with one another. Although discrete mathematics is a wide and varied field, there are certain rules that carry over into many topics. The concept of independent events and the rules of product, sum, and PIE are shared among combinatorics, set theory, and probability. In addition, De Morgan's laws are applicable in many fields of discrete mathematics.

Often, what makes discrete mathematics problems interesting and challenging are the restrictions that are placed on them. Although the field of discrete mathematics has many elegant formulas to apply, it is rare that a practical problem will fit perfectly to a specific formula. Part of the joy of discovering discrete mathematics is to learn many different approaches to problem-solving, and then be able to creatively apply disparate strategies towards a solution.

Combinatorics

Combinatorics is the mathematics of counting and arranging. Of course, most people know how to count, but combinatorics applies mathematical operations to count things that are much too large to be counted the conventional way. Combinatorics is especially useful in computer science. Combinatorics methods can be used to develop estimates about how many operations a computer algorithm will require. Combinatorics is also important for the study of discrete probability. Combinatorics methods can be used to count possible outcomes in a uniform probability experiment.

Combinatorics is often concerned with how things are arranged. In this context, an arrangement is a way objects could be grouped. The most basic rules regarding arrangements are the rule of product and the rule of sum. These rules govern how to count arrangements using the operations of multiplication and addition, respectively. A more specific type of arrangement is a permutation. A permutation is an arrangement of objects with regard to order.

A combination (not to be confused with combinatorics) is another type of arrangement that is related to permutations. A combination is an arrangement of objects without regard to order. As a field of mathematics, combinatorics is nearly as broad as discrete mathematics. Other topics within combinatorics include:

- Derangements: A permutation such that no object is in its original spot in the order.
- Rectangular grid walks: Determining the number of ways a rectangular lattice can be traversed.
- Distribution of objects into bins: Determining how objects can be grouped into bins.

Set Theory

Set theory is the branch of mathematics that is concerned about collections of objects. Sets can be discrete or continuous; discrete mathematics is primarily concerned with the former. At a basic level, set theory is concerned with how sets can be arranged, combined, and counted. The cardinality of a finite set is the number of elements in that set. For a given set A, its cardinality is denoted by $|A|$.

Example: What is the cardinality of the set of prime numbers less than 25?

The set of prime numbers less than 25 is $\{2,3,5,7,11,13,17,19,23\}$. There are 9 elements in this set, so the cardinality is 9.

Cardinality can also be extended to infinite sets. Although this kind of cardinality cannot be counted, each cardinality can be compared with another cardinality. Let A and B be sets. Their cardinalities are compared as follows:

- If there exists a bijection between A and B, then $|A|=|B|$.
- If there exists an injective function from A to B, but no bijective function, then $|A|<|B|$.

Example: Show that the set of integers and the set of even integers have the same cardinality. It might seem strange that these sets have the same cardinality. After all, the even integers are more "rare." However, these sets are both infinite. Therefore, the "common sense" thinking about finite sets must be discarded. Instead, the goal is to obtain a bijective function from the set of integers to the set of even integers:

$$f(n) = 2n, n \in \mathbb{Z}.$$

The function above gives a one-to-one correspondence between each integer n and each even integer 2n. Since the bijection is established, the set of integers and the set of even integers have the same cardinality.

A complement of a set A is the set of elements that are not in A. The study of set complements gives a number of efficient methods to calculate cardinalities of finite sets. For example, one can efficiently obtain the cardinality of a set that contains "at least one" element of another set.

The union and intersection give ways to describe how sets can be combined. De Morgan's laws give identities for the complements of unions and intersections. The principle of inclusion and exclusion, or PIE, gives a method to find the union or intersection of more than two sets.

Graph Theory

Graph theory is the study of graphs, which are a collection of connected nodes. Graphs are useful for representing all kinds of real-world problems.

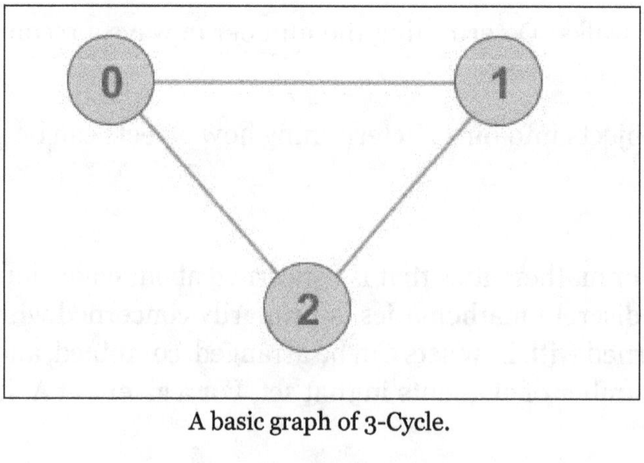

A basic graph of 3-Cycle.

Probability

A probability is a number, between 0 and 1 inclusive that represents the likelihood of an event. Discrete probability is a probability based on discrete sets of outcomes. The most basic type of probability is a uniform probability. If each outcome in a set is equally likely, then the probability of an event is equal to a ratio of cardinalities.

Let S be a sample space of outcomes. If each outcome in this set is equally likely, then the probability of an event A in S is,

$$P(A) = \frac{|A|}{|S|}.$$

Many of the rules of probability are analogous to the rules of combinatorics. The probabilistic rules of product, sum, and complement work similarly to those same rules from combinatorics. In addition, the structure of the probabilistic principle of inclusion and exclusion is the same as PIE for sets.

A discrete probability distribution is a function that takes a numerical outcome as an argument and gives a probability as a result. Discrete probability distributions can be created using the rules and guidelines described above. There are also some discrete probability distributions that show up in many problems:

- Geometric Distribution: Given repeated trials in which the probability of success is the same each time, this gives the probability that the first success will occur on a certain trial. Example: You roll a dice until you roll a 6. What is the probability that the first 6 will occur on the third roll?
- Binomial Distribution: Given a certain number of trials in which the probability of success is the same each time, this gives the probability of a certain number of successes. Example: You flip a coin 10 times. What is the probability that there will be exactly 5 heads?
- Poisson Distribution: Given a time period in which an event occurs a certain average number of times, this gives the probability that the event will occur a specific number of times. Example: A fast food drive-through gets 3 customers per minute. What's the probability they will get 4 customers in the next minute?

Although basic probabilities are based on discrete sets, the concept of probability can be extended to continuous sets by using concepts from calculus.

Statistics

A statistic is a number used to describe a set of data or a probability distribution. Statistics is widely used in many fields outside of mathematics, from biology to politics to sports. The power of statistics lies in taking a massive, varied set of data and making sense out of it. Furthermore, statistics has the power to quantify confidence in those findings. Of course, the usefulness of statistics is not without controversy, but an understanding of its theoretical underpinnings can help one avoid its misuse.

One major kind of statistic is a measure of central tendency. A measure of central tendency is a number which describes what a value of a probability distribution or data set will tend to. An expected value is the theoretical long-run average outcome of a probability experiment when it is performed many times. Somewhat related to the expected value is the mean. The mean is the average value of a set of numerical data.

Another major kind of statistic is a measure of variation. A measure of variation is a number which describes the distribution of a probability distribution or data set. The standard deviation of a probability distribution is a number that represents how much the outcomes differ from the expected value. Likewise, the standard deviation of a data set is a number that represents how much the elements of the set differ from the mean.

Although discrete statistics are based off of discrete events and probability distributions, these same concepts can be extended to continuous events and probability distributions using concepts from calculus.

Bijections

A bijection is a relationship between two sets such that each element in a set is paired with exactly one element in the other set, and vice versa. Bijections can be applied to problem solving by establishing a bijection between a set that is difficult to enumerate and a discrete stucture that is well understood. By establishing a bijection, one can take advantage of the known formulas and theorems that the discrete structure affords.

Example: The three Molloy siblings, April, Bradley, and Clark, have integer ages that sum to 15. How many possible distributions of ages are there? It is possible that an age can be 0, which means that the child was just born.

One can establish a bijection between the set of distributions of ages and a set of combinations. Consider the arrangement of stars and bars below:

$$\star\star \mid \star\star\star\star \mid \star\star\star\star\star\star\star\star\star$$

This arrangement corresponds to the following distribution of ages: April - 2, Bradley - 4, Clark - 9. There are 15 stars and 2 bars in the arrangement above. This gives a total of 17 objects, 2 of which are bars. Placing the bars in different spots among the 17 placements will give a new distribution of ages. Thus, a bijection can be established between the set of distributions of ages and the set of combinations of 2 objects out of 17.

The number of distributions of ages is,

$$\binom{17}{2} = 136.$$

Logic

A proposition is a statement that can either be true or false. Propositional logic aims to outline the rules of how these statements can be altered and combined. Similarly, Boolean algebra outlines the operations defined on variables that can take the values of true (1) or false (0). Boolean algebra is used to design computer circuits through logic gates, which take signal(s) as inputs and return a signal as an output.

Proofs in Mathematics

Proof may be what best distinguishes mathematics from other disciplines, even the sciences, which are logical, rigorous and to a greater or lesser degree (depending on the discipline) based on mathematics. By using rigorous, logically correct reasoning, we aim to prove mathematical theorems—that is, to demonstrate that something is true beyond all doubt.

It is impossible to give a formula or algorithm for proving any and all mathematical statements, yet certain approaches or strategies appear over and over in successful proofs, so studying proof itself is worthwhile. Of course, even if the subject is proof itself, we need to prove something, here we begin our study of number theory, that is, the properties of the integers (often, but not always, the non-negative integers).

The idea of proof is central to all branches of mathematics; we concentrate on proofs involving the integers for two reasons. First, it is a very good subject in which to learn to write proofs. The proofs in number theory are typically very clean and clear; there is little in the way of abstraction to cloud one's understanding of the essential points of an argument. Secondly, the integers have a central position in mathematics and are used extensively in other fields such as computer science. Although the great twentieth century mathematician G. H. Hardy boasted that he did number theory because there was no chance that it could be construed as applied mathematics, it has in fact become enormously useful and important in the study of computation and particularly in cryptography. We also find number theory intrinsically interesting, one of the most beautiful subjects in modern mathematics, and all the more interesting because of its roots in antiquity. Unless otherwise specified, then, the universe of discourse is the set of integers, $\mathbb{Z}$.

Theorem: There are infinitely many primes.

Proof: Suppose this were not the case. That is, suppose there are only finitely many primes. Then there must be a last, largest prime, call it p. Consider the number,

$$N = p! + 1 = \left(p \cdot (p-1) \cdots 3 \cdot 2 \cdot 1\right) + 1.$$

Now N is certainly larger than p. Also, N is not divisible by any number less than or equal to p, since every number less than or equal to p divides p!. Thus the prime factorization of N contains prime numbers (possibly just N itself) all greater than p. So p is not the largest prime, a contradiction. Therefore there are infinitely many primes.

This proof is an example of a proof by contradiction, one of the standard styles of mathematical proof. First and foremost, the proof is an argument. It contains sequence of statements, the last being the conclusion which follows from the previous statements. The argument is valid so the conclusion must be true if the premises are true. Let's go through the proof line by line.

- Suppose there are only finitely many primes. [This is a premise. Note the use of "suppose"].
- There must be a largest prime, call it p. [Follows from line 1, by the definition of "finitely many"].
- Let N = p! +1. [Basically just notation, although this is the inspired part of the proof; looking at p! + 1 is the key insight].
- N is larger than p. [By the definition of p!].
- N is not divisible by any number less than or equal to p. [By definition, p! is divisible by each number less than or equal to p, so p! + 1 is not].
- The prime factorization of N contains prime numbers greater than p. [Since N is divisible by each prime number in the prime factorization of N, and by line 5].
- Therefore p is not the largest prime. [By line 6, N is divisible by a prime larger than p].
- This is a contradiction. [From line 2 and line 7: the largest prime is p and there is a prime larger than p].
- Therefore there are infinitely many primes. [From line 1 and line 8: our only premise lead to a contradiction, so the premise is false].

Up through line 8, we have a valid argument with the premise "there are only finitely many primes" and the conclusion "there is a prime larger than the largest prime." This is a valid argument as each line follows from previous lines. So if the premises are true, then the conclusion must be true. However, the conclusion is NOT true. The only way out: the premise must be false.

The sort of line-by-line analysis we did above is a great way to really understand what is going on. Whenever you come across a proof, you really should make sure you understand what each line is saying and why it is true. Additionally, it is equally important to understand the overall structure of the proof. This is where using tools from logic is helpful. Luckily there are a relatively small number of standard proof styles that keep showing up again and again. Being familiar with these can help understand proof, as well as give ideas of how to write your own.

Direct Proof

The simplest (from a logic perspective) style of proof is a direct proof. Often all that is required to prove something is a systematic explanation of what everything means. Direct proofs are especially useful when proving implications. The general format to prove $P \rightarrow Q$ is this:

Assume P. Explain, explain, ..., explain. Therefore Q.

Often we want to prove universal statements, perhaps of the form $\forall x(P(x) \rightarrow Q(x))$. Again, we will want to assume P(x) is true and deduce Q(x). But what about the x? We want this to work for all x. We accomplish this by fixing x to be an arbitrary element (of the sort we are interested in).

Here are a few examples. First, we will set up the proof structure for a direct proof, then fill in the details.

Example: Prove - For all integers n, if n is even, then n^2 is even.

Solution: The format of the proof will be this: Let n be an arbitrary integer. Assume that n is even. Therefore n^2 is even. We will basically just explain what it means for n to be even, and then see what that means for n^2. Here is a complete proof.

Proof: Let n be an arbitrary integer. Suppose n is even. Then n = 2k for some integer k. Now $n^2 = (2k)^2 = 4k^2 = 2(2k^2)$. Since $2k^2$ is an integer, n^2 is even.

Proof by Contrapositive

Recall that an implication $P \rightarrow Q$ is logically equivalent to its contrapositive $\neg Q \rightarrow \neg P$. There are plenty of examples of statements which are hard to prove directly, but whose contrapositive can easily be proved directly. This is all that proof by contrapositive does. It gives a direct proof of the contrapositive of the implication. This is enough because the contrapositive is logically equivalent to the original implication.

The skeleton of the proof of $P \rightarrow Q$ by contrapositive will always look roughly like this:

Assume $\neg Q$. Explain, explain, ... explain. Therefore $\neg P$.

If there are variables and quantifiers, we set them to be arbitrary elements of our domain.

Example: Is the statement "for all integers n, if n^2 is even, then n is even" true?

Solution: This is the converse of the statement we proved above using a direct proof. From trying a few examples, this statement definitely appears this is true. So let's prove it.

A direct proof of this statement would require fixing an arbitrary n and assuming that n^2 is even. But it is not at all clear how this would allow us to conclude anything about n. Just because n^2 = 2k does not in itself suggest how we could write n as a multiple of 2. Write the contrapositive of the statement. We get, for all integers n, if n is odd then n^2 is odd. This looks much more promising.

Let n be an arbitrary integer. Suppose that n is not even. This means that in other words this is the same as saying. Therefore n^2 is not even.

Proof: We will prove the contrapositive. Let n be an arbitrary integer. Suppose that n is not even, and thus odd. Then $n = 2k+1$ for some integer k. Now $n^2 = (2k+1)^2 = 4k^2 + 4k + 1 = 2(2k^2 + 2k) + 1$. Since $2k^2 + 2k$ is an integer, we see that n^2 is odd and therefore not even.

Example: Prove - For all integers a and b, if a + b is odd, then a is odd or bb is odd.

Solution: The problem with trying a direct proof is that it will be hard to separate a and b from knowing something about a + b. On the other hand, if we know something about a and b separately, then combining them might give us information about a + b. The contrapositive of the statement we are trying to prove is: for all integers a and b, if a and b are even, then a + b is even. Thus our proof will have the following format:

Let a and b be integers. Assume that a and bb are both even. Therefore a + b is even.

Proof: Let a and b be integers. Assume that a and b are even. Then a = 2k and b = 2l for some integers k and l. Now $a+b=2k+2l=2(k+1)$. Since k + l is an integer, we see that a + b is even, completing the proof. Our assumption that a and b are even is really the negation of a or b is odd. We used De Morgan's law here.

We have seen how to prove some statements in the form of implications: either directly or by contrapositive. Some statements are not written as implications to begin with.

Proof by Contradiction

There might be statements which really cannot be rephrased as implications. For example, " $\sqrt{2}$ is irrational." In this case, it is hard to know where to start. What can we assume? Well, say we want to prove the statement P. What if we could prove that ¬P→Q where Q was false? If this implication is true, and Q is false, what can we say about ¬P? It must be false as well, which makes P true.

This is why proof by contradiction works. If we can prove that ¬P leads to a contradiction, then the only conclusion is that ¬P is false, so P is true. That's what we wanted to prove. In other words, if it is impossible for P to be false, P must be true.

Example: Prove that $\sqrt{2}$ is irrational.

Solution: Proof - Suppose not. Then $\sqrt{2}$ is equal to a fraction $\frac{a}{b}$. Without loss of generality, assume $\frac{a}{b}$ is in lowest terms (otherwise reduce the fraction). So,

$$2=\frac{a^2}{b^2}$$

$$2b^2=a^2$$

Thus a^2 is even, and as such a is even. So a = 2k for some integer k, and $a^2 = 4k^2$. We then have,

$$2b^2=4k^2$$

$$b^2=2k^2$$

Thus b^2 is even, and as such b is even. Since a is also even, we see that $\frac{a}{b}$ is not in lowest terms, a contradiction. Thus $\sqrt{2}$ is irrational.

Proof by Counter Example

It is almost never okay to prove a statement with just an example. Certainly none of the statements proved above can be proved through an example. This is because in each of those cases we are trying to prove that something holds of all integers. We claim that n^2 being even implies that n is even, no matter what integer n we pick. Showing that this works for n = 4 is not even close to enough. This cannot be stressed enough. If you are trying to prove a statement of the form ∀xP(x), you absolutely cannot prove this with an example.

However, existential statements can be proven this way. If we want to prove that there is an integer n such that $n^2 - n + 4$ is not prime all we need to do is find one. This might seem like a silly thing to want to prove until you try a few values for n.

n	1	2	3	4	5	6	7
$n^2 - n + 41$	41	43	47	53	61	71	83

So far we have gotten only primes. You might be tempted to conjecture, "For all positive integers n, the number $n^2 - n + 41$ is prime." If you wanted to prove this, you would need to use a direct proof, a proof by contrapositive, or another style of proof. In fact, we can prove this conjecture is false by proving its negation: "There is a positive integer n such that $n^2 - n + 41$ is not prime." Since this is an existential statement, it suffices to show that there does indeed exist such a number.

In fact, we can quickly see that $n = 41$ will give 41^2 which is certainly not prime. You might say that this is a counterexample to the conjecture that $n^2 - n + 41$ is always prime. Since so many statements in mathematics are universal, making their negations existential, we can often prove that a statement is false (if it is) by providing a counterexample.

Example: Above we proved, "for all integers a and b, if a + b is odd, then a is odd or bb is odd." Is the converse true?

Solution: The converse is the statement, "for all integers a and b, if a is odd or b is odd, then a + b is odd." This is false! How do we prove it is false? We need to prove the negation of the converse. Let's look at the symbols. The converse is,

$$\forall a \forall b\big((O(a) \vee O(b)) \rightarrow O(a+b)\big).$$

We want to prove the negation:

$$\neg \forall a \forall b\big((O(a) \vee O(b)) \rightarrow O(a+b)\big).$$

Simplify using the rules from above:

$$\exists a \exists b\big((O(a) \vee O(b)) \wedge \neg O(a+b)\big).$$

As the negation passed by the quantifiers, they changed from $\forall$ to $\exists$. We then needed to take the negation of an implication, which is equivalent to asserting the if part and not the then part.

To prove that the converse is false we need to find two integers a and bb so that a is odd or b is odd, but a + b is not odd (so even). That's easy: 1 and 3. (remember, "or" means one or the other or both). Both of these are odd, but $1 + 3 = 4$ is not odd.

Proof by Cases

We could go on and on and on about different proof styles, but instead we will end with one final useful technique: proof by cases. The idea is to prove that PP is true by proving that $Q \rightarrow P$ and $\neg Q \rightarrow P$ for some statement Q. So no matter what, whether or not Q is true, we know that P is true. In fact, we could generalize this. Suppose we want to prove P. We know that at least one of the statements $Q_1, Q_2, \ldots, Q_n$ are true. If we can show that $Q_1 \rightarrow P$ and $Q_2 \rightarrow P$ and so on all the way to $Q_n \rightarrow P$, then we can conclude P. The key thing is that we want to be sure that one of our cases (the Q_i's) must be true no matter what.

Example: Prove - For any integer n, the number $(n^3 - n)$ is even.

Solution: It is hard to know where to start this, because we don't know much of anything about n. We might be able to prove that $n^3 - n$ is even if we knew that n was even. In fact, we could probably prove that $n^3 - n$ was even if n was odd. But since n must either be even or odd, this will be enough. Here's the proof.

Proof: We consider two cases: if n is even or if n is odd.

Case 1: n is even. Then n = 2k for some integer k. This give,

$$\begin{aligned} n^3 - n &= 8k^3 - 2k \\ &= 2(4k^2 - k), \end{aligned}$$

and since $4k^2 - k$ is an integer, this says that $n^3 - n$ is even.

Case 2: n is odd. Then n = 2k + 1 for some integer k. This gives,

$$\begin{aligned} n^3 - n &= (2k+1)^3 - (2k+1) \\ &= 8k^3 + 6k^2 + 6k + 1 - 2k - 1 \\ &= 2(4k^3 + 3k^2 + 2k), \end{aligned}$$

and since $4k^3 + 3k^2 + 2k$ is an integer, we see that $n^3 - n$ is even again. Since $n^3 - n$ is even in both exhaustive cases, we see that $n^3 - n$ is indeed always even.

Probability Theory

Probability theory was invented in the 17th century by two French mathematicians, Blaise Pascal and Pierre de Fermat, who were dealing with mathematical problems regarding of chance. Let us get the concept of some definitions:

- Random Experiment: An experiment in which all possible outcomes are known and the exact output cannot be predicted in advance is called a random experiment. Tossing a fair coin is an example of random experiment.
- Sample Space: When we perform an experiment, then the set S of all possible outcomes is called the sample space. If we toss a coin, the sample space S = {H,T}.
- Event: Any subset of a sample space is called an event. After tossing a coin, getting Head on the top is an event.

The word "probability" means the chance of occurrence of a particular event. The best we can say is how likely they are to happen, using the idea of probability.

$$\text{Probability of occurence of an event} = \frac{\text{Total number of favourable outcome}}{\text{Total number of Outcomes}}$$

As the occurrence of any event varies between 0% and 100%, the probability varies between 0 and 1.

Steps to Find the Probability

- Step 1: Calculate all possible outcomes of the experiment.
- Step 2: Calculate the number of favorable outcomes of the experiment.
- Step 3: Apply the corresponding probability formula.

Tossing a Coin

If a coin is tossed, there are two possible outcomes – Heads (H) or Tails (T). So, Total number of outcomes = 2. Hence, the probability of getting a Head (H) on top is 1/2 and the probability of getting a Tails (T) on top is 1/2.

Throwing a Dice

When a dice is thrown, six possible outcomes can be on the top – 1, 2, 3, 4, 5, 6.

- The probability of any one of the numbers is 1/6.
- The probability of getting even numbers is 3/6 = 1/2.
- The probability of getting odd numbers is 3/6 = 1/2.

Taking Cards from a Deck

From a deck of 52 cards, if one card is picked find the probability of an ace being drawn and also find the probability of a diamond being drawn.

- Total number of possible outcomes is 52.
- Outcomes of being an ace is 4.
- Probability of being an ace = 4/52 = 1/13.
- Probability of being a diamond = 13/52 = 1/4.

Probability Axioms

- The probability of an event always varies from 0 to 1. $[0 \leq P(x) \leq 1]$.
- For an impossible event the probability is 0 and for a certain event the probability is 1.

- If the occurrence of one event is not influenced by another event, they are called mutually exclusive or disjoint.
- If $A_1, A_2 .... A_n$ are mutually exclusive/disjoint events, then $P(A_i \cap A_j) = \phi$ for $i \neq j$ and $P(A_1 \cup A_2 \cup .... A_n) = P(A_1) + P(A_2) + ... P(A_n)$.

Properties of Probability

- If there are two events x and $\bar{x}$ which are complementary, then the probability of the complementary event is – $p(\bar{x}) = 1 - p(x)$.
- For two non-disjoint events A and B, the probability of the union of two events: $P(A \cup B) = P(A) + P(B)$.
- If an event A is a subset of another event B (i.e. $A \subset B$), then the probability of A is less than or equal to the probability of B. Hence, $A \subset B$ implies $P(A) \leq p(B)$.

Conditional Probability

The conditional probability of an event B is the probability that the event will occur given an event A has already occurred. This is written as P(B|A).

Mathematically: $P(B|A) = P(A \cap B)/P(A)$.

If event A and B are mutually exclusive, then the conditional probability of event B after the event A will be the probability of event B that is P(B).

Example: In a country 50% of all teenagers own a cycle and 30% of all teenagers own a bike and cycle. What is the probability that a teenager owns bike given that the teenager owns a cycle?

Solution: Let us assume A is the event of teenagers owning only a cycle and B is the event of teenagers owning only a bike.

So, $P(A) = 50/100 = 0.5$ and $P(A \cap B) = 30/100 = 0.3$ from the given problem.

$$P(B|A) = P(A \cap B)/P(A) = 0.3/0.5 = 0.6$$

Hence, the probability that a teenager owns bike given that the teenager owns a cycle is 60%.

Example: In a class, 50% of all students play cricket and 25% of all students play cricket and volleyball. What is the probability that a student plays volleyball given that the student plays cricket?

Solution: Let us assume A is the event of students playing only cricket and B is the event of students playing only volleyball.

So, $P(A) = 50/100 = 0.5$ and $P(A \cap B) = 25/100 = 0.25$ from the given problem.

$$P(B|A) = P(A \cap B)/P(A) = 0.25/0.5 = 0.5$$

Hence, the probability that a student plays volleyball given that the student plays cricket is 50%.

Bayes' Theorem

Theorem: If A and B are two mutually exclusive events, where P(A) is the probability of A and P(B) is the probability of B, P(A|B) is the probability of A given that B is true. P(B|A) is the probability of B given that A is true, then Bayes' Theorem states:

$$P(A|B)=\frac{P(B|A)P(A)}{\sum_{i=1}^{n}P(B|A_i)P(A_i)}$$

Applications of Bayes' Theorem

- In situations where all the events of sample space are mutually exclusive events.
- In situations where either $P(A_i \cap B)$ for each A_i or $P(A_i)$ and $P(B|A_i)$ for each A_i is known.

Example: Consider three pen-stands. The first pen-stand contains 2 red pens and 3 blue pens; the second one has 3 red pens and 2 blue pens; and the third one has 4 red pens and 1 blue pen. There is equal probability of each pen-stand to be selected. If one pen is drawn at random, what is the probability that it is a red pen?

Solution: Let A_i be the event that i^{th} pen-stand is selected. Here, i = 1,2,3. Since probability for choosing a pen-stand is equal, $P(A_i)=1/3$.

Let B be the event that a red pen is drawn. The probability that a red pen is chosen among the five pens of the first pen-stand,

$$P(B|A_1)=2/5$$

The probability that a red pen is chosen among the five pens of the second pen-stand,

$$P(B|A_2)=3/5$$

The probability that a red pen is chosen among the five pens of the third pen-stand,

$$P(B|A_3)=4/5$$

According to Bayes' Theorem,

$$\begin{aligned}P(B)&=P(A_1).P(B|A_1)+P(A_2).P(B|A_2)+P(A_3).P(B|A_3)\\&=1/3.2/5+1/3.3/5+1/3.4/5\\&=3/5\end{aligned}$$

Here are some typical statements that you might see concerning probability:

- The chance of getting a flush in a 5-card poker hand is about 2 in 1000.

- The chance that this randomized primality testing algorithm outputs "prime" when the input is not prime is at most one in a trillion.
- In this load-balancing scheme, the probability that any processor has to deal with more than 12 requests is negligible.
- The average time between system failures is about 3 days.
- There is a 30% chance of a magnitude 8.0 earthquake in Northern California before 2030.

Implicit in all such statements is the notion of an underlying probability space. This may be the result of a random experiment that we have ourselves constructed, or some model we build of the real world. None of these statements makes sense unless we specify the probability space we are talking about: for this reason, statements like 5 (which are typically made without this context) are almost content-free.

Let us try to understand all this more clearly. The first important notion here is that of a random experiment. An example of such an experiment is tossing a coin 4 times, or dealing a poker hand. In the first case an outcome of the experiment might be HTHT or it might be HHHT. The question we are interested in might be "what is the chance that there are exactly two H's?" Well, the number of outcomes that meet this condition is $\binom{4}{2} = \frac{4!}{2!2!} = 6$ (corresponding to choosing the positions of the two H's in the sequence of four tosses); these outcomes are HHTT, HTHT, HTTH, THHT, THTH, TTHH. On the other hand, the total number of distinct outcomes for this experiment is $2^4 = 16$. If the coin is fair then all these 16 outcomes are equally likely, so the chance that there are exactly two H's is 6/16 = 3/8.

As we saw with counting, there is a common framework in which we can view random experiments about flipping coins, dealing cards, rolling dice, etc. A finite process is the following: We are given a finite population U, of cardinality n. In the case of coin tossing, U = {H, T}, and in card dealing, U is the set of 52 cards.

An experiment consists of drawing a sample of k elements from U (where we can consider either sampling with or without replacement). The outcome of the random experiment is called a sample point. Thus HTHT is an example of a sample point. The sample space, often denoted by Ω, is the set of all possible outcomes. In our example the sample space has 16 elements.

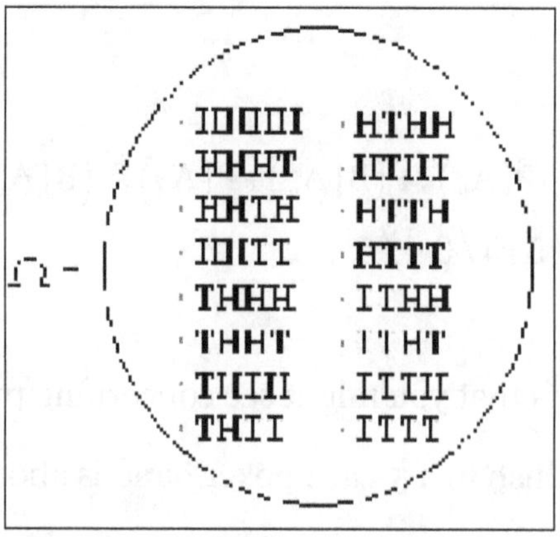

A probability space is a sample space Ω, together with a probability $Pr[\omega]$ for each sample point ω, such that:

- $0 \le Pr[\omega] \le 1$ for all $\omega \in \Omega$.
- $\sum_{\omega\in\Omega} Pr[\omega] = 1$, i.e., the sum of the probabilities of all outcomes is 1.

The easiest way to assign probabilities to sample points is uniformly (in the coin tossing example): if $|\Omega| = N$, then $Pr[x] = \frac{1}{N}\ \forall x \in \Omega$. We will see examples of non-uniform probability distributions soon.

Here's another example: dealing a poker hand. In this case, our sample space Ω = {all possible poker hands}, which corresponds to choosing k = 5 objects without replacement from a set of size n = 52 where order does not matter. Hence, as we saw in the previous Note, $|\Omega| = \binom{52}{5} = \frac{52\times51\times50\times49\times48}{5\times4\times3\times2\times1} = 2{,}598{,}960$. Since the probability of each outcome is equally likely, this implies that the probability of any particular hand, such as, {5♥, 3♣, 7♠, 8♣, K♥}, is $\frac{1}{2{,}598{,}960}$.

As we saw in the coin tossing example above, what we are often interested in knowing after performing an experiment is whether a certain event occurred. Thus we considered the event that there were exactly two H's in the four tosses of the coin. Here are some more examples of events we might be interested in:

- Sum of the rolls of 2 dice is ≥ 10.
- Poker hand is a flush (i.e., all 5 cards have the same suit).
- In n coin tosses, at least $\ge \frac{n}{3}$ of the tosses come up tails.

Let us now formalize the notion of an event. Formally, an event A is just a subset of the sample space, $A \subseteq \Omega$. The event "exactly 2 H's in four tosses of the coin" is the subset: {HHTT, HTHT, HTTH, THHT, THTH, TTHH} $\subseteq \Omega$.

How should we define the probability of an event A? Naturally, we should just add up the probabilities of the sample points in A. For any event $A \subseteq \Omega$, we define the probability of A to be,

$$Pr[A] = \sum_{\omega\in A} Pr[\omega].$$

Thus the probability of getting exactly two H's in four coin tosses can be calculated using this definition as follows. A consists of all sequences that have exactly two H's, and so $|A| = \binom{4}{2} = 6$. For this example, there are $2^4 = 16$ possible outcomes for flipping four coins. Thus, each sample point $\omega \in$ A has probability $\frac{1}{16}$; and, there are six sample points in A, giving us $Pr[A] = 6\cdot\frac{1}{16} = \frac{3}{8}$.

We will now look at examples of probability spaces and typical events that may occur in such experiments.

1. Flip a fair coin. Here $\Omega = \{H,T\}$, and $\Pr[H] = \Pr[T] = \frac{1}{2}$.

2. Flip a fair coin three times. Here $\Omega = \{(t_1,t_2,t_3) : t_i \in \{H,T\}\}$, where t_i gives the outcome of the ith toss. Thus Ω consists of $2^3 = 8$ points, each with equal probability $\frac{1}{8}$. More generally, if we flip the coin n times, we get a sample space of size 2^n (corresponding to all words of length n over the alphabet {H,T}), each point having probability $\frac{1}{2^n}$. We can look, e.g., at the event A that all three coin tosses are the same. Then A = {HHH, TTT}, with each sample point having probability $\frac{1}{8}$. Thus, $\Pr[A] = \Pr[HHH] + \Pr[T\ T\ T] = \frac{1}{8} + \frac{1}{8} = \frac{1}{4}$.

3. Flip a biased coin three times. Suppose the bias is two-to-one in favor of Heads, i.e., it comes up Heads with probability $\frac{2}{3}$ and Tails with probability $\frac{1}{3}$. The sample space here is exactly the same as in the previous example. However, the probabilities are different. For example, $\Pr[HHH] = \frac{2}{3} \times \frac{2}{3} \times \frac{2}{3} = \frac{8}{27}$, while $\Pr[THH] = \frac{1}{3} \times \frac{2}{3} \times \frac{2}{3} = \frac{4}{27}$. More generally, if we flip a biased coin with Heads probability p (and Tails probability 1−p) n times, the probability of a given sequence is $p^r(1-p)^{n-r}$, where r is the number of H's in the sequence. Let A be the same event as in the previous example. Then $\Pr[A] = \Pr[HHH] + \Pr[TTT] = \frac{8}{27} + \frac{1}{27} = \frac{9}{27} = \frac{1}{3}$. As a second example, let B be the event that there are exactly two Heads. We know that the probability of any outcome with two Heads (and therefore one Tail) is $\left(\frac{2}{3}\right)^2 \times \left(\frac{1}{3}\right) = \frac{4}{27}$. How many such outcomes are there? Well, there are $\binom{3}{2} = 3$ ways of choosing the positions of the Heads, and these choices completely specify the sequence. So $\Pr[B] = 3 \times \frac{4}{27} = \frac{4}{9}$. More generally, the probability of getting exactly r Heads from n tosses of a biased coin with Heads probability p is $\binom{n}{r} p^r (1-p)^{n-r}$. Biased coin-tossing sequences show up in many contexts: for example, they might model the behavior of n trials of a faulty system, which fails each time with probability p.

4. Roll two dice: Then $\Omega = \{(i, j) : 1 \leq i, j \leq 6\}$. Each of the 36 outcomes has equal probability, $\frac{1}{36}$. We can look at the event A that the sum of the dice is at least 10, and B the event that there is at

least one 6. In this example, our probability space is uniform, i.e., all the sample points have the same probability (which must be $\frac{1}{|\Omega|}$, where $|\Omega|$ denotes the size of Ω). In such circumstances, the probability of any event A is clearly just,

$$\Pr[A] = \frac{\text{\# of sample points in A}}{\text{\# of sample points in }\Omega} = \frac{|A|}{|\Omega|}.$$

So for uniform spaces, computing probabilities reduces to counting sample points! Using this observation, it is now easy to compute the probabilities of the two events A and B above: $\Pr[A] = \frac{6}{36} = \frac{1}{6}$, and $\Pr[B] = \frac{11}{36}$.

5. Card Shuffling: Shuffle a deck of cards. Here Ω consists of the 52! Permutations of the deck, each with equal probability $\frac{1}{52!}$. [We're really talking about an idealized mathematical model of shuffling here; in real life, there will always be a bit of bias in our shuffling. However, the mathematical model is close enough to be useful].

6. Poker Hands: Shuffle a deck of cards, and then deal a poker hand. Here Ω consists of all possible five-card hands, each with equal probability (because the deck is assumed to be randomly shuffled). As we saw above, the number of such hands is $\binom{52}{5}$. What is the probability that our poker hand is a flush? [For those who are not addicted to gambling, a flush is a hand in which all cards have the same suit, say Hearts]. To compute this probability, we just need to figure out how many poker hands are flushes. Well, there are 13 cards in each suit, so the number of flushes in each suit is $\binom{13}{5}$. The total number of flushes is therefore $4 \cdot \binom{13}{5}$. So we have,

$$\Pr[\text{hand is a flush}] = \frac{4 \cdot \binom{13}{5}}{\binom{52}{5}} = \frac{4 \cdot 13! \cdot 5! \cdot 47!}{5! \cdot 8! \cdot 52!} = \frac{4 \cdot 13 \cdot 12 \cdot 11 \cdot 10 \cdot 9}{52 \cdot 51 \cdot 50 \cdot 49 \cdot 48} \approx 0.002.$$

7. Balls and Bins: Throw 20 (labeled) balls into 10 (labeled) bins, so that each ball is equally likely to land in any bin, regardless of what happens to the other balls. (Thus, in the terminology of the previous Note, we are in the situation of "sampling with replacement" and order does matter.) Here $\Omega = \{(b_1, b_2, \ldots, b_{20}) : 1 \le b_i \le 10\}$; the component b_i denotes the bin in which ball i lands. There are 10^{20} possible outcomes (why?), each with probability $\frac{1}{10^{20}}$. More generally, if we throw m balls into n bins, we have a sample space of size n^m. [above is the special case with m = 3 and n = 2, and example 4 is the special case m = 2, n = 6]. Let A be the event that bin 1 is empty. Again, we just need to count how many outcomes have this property. And this is exactly the number of ways all 20 balls

can fall into the remaining nine bins, which is 9^{20}. Hence $\Pr[A]=\frac{9^{20}}{10^{20}}=\left(\frac{9}{10}\right)^{20}\approx 0.12$. What is the probability that bin 1 contains at least one ball? This is easy: this event, call it $\bar{A}$, is the complement of A, i.e., it consists of precisely those sample points that are not in A. So $\Pr[\bar{A}]=1-\Pr[A]\approx 0.88$. More generally, if we throw m balls into n bins, we have,

$$\Pr[\text{bin 1 is empty}]=\left(\frac{n-1}{n}\right)^{m}=\left(1-\frac{1}{n}\right)^{m}.$$

Balls and bins is another probability space that shows up very often in Computer Science for example, we can think of it as modeling a load balancing scheme, in which each job is sent to a random processor.

Birthday Paradox

The "birthday paradox" is a remarkable phenomenon that examines the chances that two people in a group have the same birthday. It is a "paradox" not because of a logical contradiction, but because it goes against intuition. For ease of calculation, we take the number of days in a year to be 365. If we consider the case where there are n people in a room, then $|\Omega|=365^{n}$. Let A = "At least two people have the same birthday," and let $\bar{A}$ = "No two people have the same birthday." It is clear that $\Pr[A]=1-\Pr[\bar{A}]$. We will calculate $\Pr[\bar{A}]$, since it is easier, and then find out Pr[A]. How many ways are there for no two people to have the same birthday? Well, there are 365 choices for the first person, 364 for the second, . . . , 365−n+1 choices for the nth person, for a total of 365 × 364 × ··· × (365−n+1). (this is just sampling without replacement with 365 bins and n balls; as we saw in the previous Note, the number of outcomes is $\frac{365!}{(365-n)!}$, which is what we just got.) Thus we have $\Pr[\bar{A}]=\frac{|\bar{A}|}{|\Omega|}=\frac{365\times 364\times\cdots\times(365-n+1)}{356^{n}}$. Then $\Pr[A]=1-=\frac{365\times 364\times\cdots\times(365-n+1)}{356^{n}}$.

This allows us to compute Pr[A] as a function of the number of people, n. Of course, as n increases Pr[A] increases. In fact, with n = 23 people you should be willing to bet that at least two people do have the same birthday, since then Pr[A] is larger than 50%! For n = 60 people, Pr[A] is over 99%.

Counting Theory

Counting mainly encompasses fundamental counting rule, the permutation rule, and the combination rule.

The Rules of Sum and Product

The Rule of Sum and Rule of Product are used to decompose difficult counting problems into simple problems.

- The Rule of Sum: If a sequence of tasks $T_1, T_2, \ldots, T_m$ can be done in $w_1, w_2, \ldots w_m$ ways respectively (the condition is that no tasks can be performed simultaneously), then the number of ways to do one of these tasks is $w_1 + w_2 + \ldots + w_m$. If we consider two tasks A and B which are disjoint (i.e. $A \cap B = \emptyset$), then mathematically $|A \cup B| = |A| + |B|$.
- The Rule of Product: If a sequence of tasks $T_1, T_2, \ldots, T_m$ can be done in $w_1, w_2, \ldots w_m$ ways respectively and every task arrives after the occurrence of the previous task, then there are $w_1 \times w_2 \times \ldots \times w_m$ ways to perform the tasks. Mathematically, if a task B arrives after a task A, then $|A \times B| = |A| \times |B|$.

Example: A boy lives at X and wants to go to School at Z. From his home X he has to first reach Y and then Y to Z. He may go X to Y by either 3 bus routes or 2 train routes. From there, he can either choose 4 bus routes or 5 train routes to reach Z. How many ways are there to go from X to Z?

Solution: From X to Y, he can go in 3 + 2 = 5 ways (Rule of Sum). Thereafter, he can go Y to Z in 4 + 5 = 9 ways (Rule of Sum). Hence from X to Z he can go in 5 × 9 = 45 ways (Rule of Product).

Permutations

A permutation is an arrangement of some elements in which order matters. In other words a Permutation is an ordered Combination of elements.

Examples:

- From a set S = {x, y, z} by taking two at a time, all permutations are: xy, yx, xz, zx, yz, zy.
- We have to form a permutation of three digit numbers from a set of numbers S = {1, 2, 3}. Different three digit numbers will be formed when we arrange the digits. The permutation will be = 123, 132, 213, 231, 312, 321.

Number of Permutations

The number of permutations of 'n' different things taken 'r' at a time is denoted by nP_r,

$$nP_r = \frac{n!}{(n-r)!}$$

where $n! = 1.2.3....(n-1).n$.

Proof: Let there be 'n' different elements. There are n number of ways to fill up the first place. After filling the first place (n-1) number of elements is left. Hence, there are (n-1) ways to fill up the second place. After filling the first and second place, (n-2) number of elements is left. Hence, there are (n-2) ways to fill up the third place. We can now generalize the number of ways to fill up r-th place as $\left[n-(r-1)\right]=n-r+1$. So, the total no. of ways to fill up from first place up to r-th-place:

$$nP_r = n(n-1)(n-2).....(n-r+1)$$
$$=\left[n(n-1)(n-2)...(n-r+1)\right]\left[(n-r)(n-r-1)...3.2.1\right]$$
$$/\left[(n-r)(n-r-1)...3.2.1\right]$$

Hence, $nP_r = n!/(n-r)!$

Important Formulas of Permutation

- If there are n elements of which a_1 are alike of some kind, a_2 are alike of another kind; a_3 are alike of third kind and so on and a_r are of rth kind, where $(a_1 + a_2 + ... a_r) = n$. Then, number of permutations of these n objects is $= n!/\left[(a_1!(a_2!)...(a_r!)\right]$.
- Number of permutations of n distinct elements taking n elements at a time $= n_{P_n} = n!$
- The number of permutations of n dissimilar elements taking r elements at a time, when x particular things always occupy definite places $= n - x_{P_{r-x}}$.
- The number of permutations of n dissimilar elements when r specified things always come together is $- r!(n-r+1)!$
- The number of permutations of n dissimilar elements when r specified things never come together is $- n! - \left[r!(n-r+1)!\right]$.
- The number of circular permutations of n different elements taken x elements at time = ${}^n p_x / x$.
- The number of circular permutations of n different things = ${}^n p_n / n$.

Examples:

1. From a bunch of 6 different cards, how many ways we can permute it?

2. Solution: As we are taking 6 cards at a time from a deck of 6 cards, the permutation will be ${}^6P_6$ = 6! = 720.

3. In how many ways can the letters of the word 'READER' be arranged?

4. Solution: There are 6 letters word (2 E, 1 A, 1D and 2R.) in the word 'READER'. The permutation will be =6!/[(2!)(1!)(1!)(2!)] = 180.

5. In how ways can the letters of the word 'ORANGE' be arranged so that the consonants occupy only the even positions?

6. Solution: There are 3 vowels and 3 consonants in the word 'ORANGE'. Number of ways of ar-

ranging the consonants among themselves $= {}^3P_3 = 3! = 6$. The remaining 3 vacant places will be filled up by 3 vowels in ${}^3P_3 = 3! = 6$ ways. Hence, the total number of permutation is $6 \times 6 = 36$.

Combinations

A combination is selection of some given elements in which order does not matter. The number of all combinations of n things, taken r at a time is,

$$ {}^nC_r = \frac{n!}{r!(n-r)!} $$

Example: Find the number of subsets of the set {1, 2, 3, 4, 5, 6} having 3 elements.

Solution: The cardinality of the set is 6 and we have to choose 3 elements from the set. Here, the ordering does not matter. Hence, the number of subsets will be ${}^6C_3 = 20$.

Example: There are 6 men and 5 women in a room. In how many ways we can choose 3 men and 2 women from the room?

Solution: The number of ways to choose 3 men from 6 men is ${}^6C_3$ and the number of ways to choose 2 women from 5 women is ${}^5C_2$. Hence, the total number of ways is – ${}^6C_3 \times {}^5C_2 = 20 \times 10 = 200$.

Example: How many ways can you choose 3 distinct groups of 3 students from total 9 students?

Solution: Let us number the groups as 1, 2 and 3. For choosing 3 students for 1st group, the number of ways – ${}^9C_3$. The number of ways for choosing 3 students for 2nd group after choosing 1st group – ${}^6C_3$. The number of ways for choosing 3 students for 3rd group after choosing 1st and 2nd group – ${}^3C_3$. Hence, the total number of ways $= {}^9C_3 \times {}^6C_3 \times {}^3C_3 = 84 \times 20 \times 1 = 1680$.

Pascal's Identity

Pascal's identity, first derived by Blaise Pascal in 17th century, states that the number of ways to choose k elements from n elements is equal to the summation of number of ways to choose (k-1) elements from (n-1) elements and the number of ways to choose elements from n-1 elements.

Mathematically, for any positive integers k and n: ${}^nC_k = {}^{n-1}C_{k-1} + {}^{n-1}C_k$.

Proof:

$$ {}^{n-1}C_{k-1} + {}^{n-1}C_k $$

$$ = \frac{(n-1)!}{(k-1)!(n-k)!} + \frac{(n-1)!}{k!(n-k-1)!} $$

$$ = (n-1)!\left(\frac{k}{k!(n-k)!} + \frac{n-k}{k!(n-k)!}\right) $$

$$ = (n-1)!\frac{n}{k!(n-k)!} = \frac{n!}{k!(n-k)!} = n_{C_k} $$

Pigeonhole Principle

In 1834, German mathematician, Peter Gustav Lejeune Dirichlet, stated a principle which he called the drawer principle. Now, it is known as the pigeonhole principle. Pigeonhole Principle states that if there are fewer pigeon holes than total number of pigeons and each pigeon are put in a pigeon hole, then there must be at least one pigeon hole with more than one pigeon. If n pigeons are put into m pigeonholes where n > m, there's a hole with more than one pigeon.

Examples:

- Ten men are in a room and they are taking part in handshakes. If each person shakes hands at least once and no man shakes the same man's hand more than once then two men took part in the same number of handshakes.
- There must be at least two people in a class of 30 whose names start with the same alphabet.

The Inclusion-Exclusion Principle

The Inclusion-exclusion principle computes the cardinal number of the union of multiple non-disjoint sets. For two sets A and B, the principle states:

$$|A \cup B| = |A| + |B| - |A \cap B|$$

For three sets A, B and C, the principle states:

$$|A \cup B \cup C| = |A| + |B| + |C| - |A \cap B| - |A \cap C| - |B \cap C| + |A \cap B \cap C|$$

The generalized formula:

$$\left|\bigcup_{i=1}^{n} A_i\right| = \sum_{1 \le i < j < k \le n} |A_i \cap A_j| + \sum_{1 \le i < j < k \le n} |A_i \cap A_j \cap A_k| - \cdots + (-1)^{\backslash n-1} |A_1 \cap A_2|$$

Example: How many integers from 1 to 50 are multiples of 2 or 3 but not both?

Solution: From 1 to 100, there are 50/2 = 25 numbers which are multiples of 2. There are 50/3=16 numbers which are multiples of 3. There are 50/6=8 numbers which are multiples of both 2 and 3.

So, $|A| = 25$, $|B| = 16$ and $|A \cap B| = 8$.

$$|A \cup B| = |A| + |B| - |A \cap B| = 25 + 16 - 8 = 33$$

Example: In a group of 50 students 24 like cold drinks and 36 like hot drinks and each student likes at least one of the two drinks. How many like both coffee and tea?

Solution: Let X be the set of students who like cold drinks and Y be the set of people who like hot drinks.

So, $|X \cup Y| = 50$, $|X| = 24$, $|Y| = 36$

$$|X \cap Y| = |X| + |Y| - |X \cup Y| = 24 + 36 - 50 = 60 - 50 = 10$$

Hence, there are 10 students who like both tea and coffee.

Propositional Logic

Propositional Logic is concerned with statements to which the truth values, "true" and "false", can be assigned. The purpose is to analyze these statements either individually or in a composite manner. A proposition is a collection of declarative statements that has either a truth value "true" or a truth value "false". A propositional consists of propositional variables and connectives. We denote the propositional variables by capital letters (A, B, etc.). The connectives connect the propositional variables.

Some examples of Propositions are given below:

- "Man is Mortal", it returns truth value "TRUE".
- "12 + 9 = 3 – 2", it returns truth value "FALSE".

The following is not a Proposition: "A is less than 2". It is because unless we give a specific value of A, we cannot say whether the statement is true or false.

Connectives

In propositional logic generally we use five connectives which are: OR (∨), AND (∧), Negation/ NOT (¬), Implication / if-then (→) and If and only if (⇔).

1. OR (∨): The OR operation of two propositions A and B (written as $A \vee B$) is true if at least any of the propositional variables A or B is true. The truth table is as follows:

A	B	$A \vee B$
True	True	True
True	False	True
False	True	True
False	False	False

2. AND (∧): The AND operation of two propositions A and B (written as $A \wedge B$) is true if both the propositional variable A and B is true. The truth table is as follows:

A	B	$A \wedge B$
True	True	True
True	False	False
False	True	False
False	False	False

3. Negation (¬): The negation of a proposition A (written as $\neg A$) is false when A is true and is true when A is false. The truth table is as follows:

A	$\neg A$
True	False
False	True

4. Implication / if-then (→): An implication $A \rightarrow B$ is the proposition "if A, then B". It is false if A is true and B is false. The rest cases are true. The truth table is as follows:

A	B	$A \rightarrow B$
True	True	True
True	False	False
False	True	True
False	False	True

5. If and only if (⇔): A⇔B is bi-conditional logical connective which is true when p and q are same, i.e. both are false or both are true. The truth table is as follows:

A	B	$A \Leftrightarrow B$
True	True	True
True	False	False
False	True	False
False	False	True

Tautology

A Tautology is a formula which is always true for every value of its propositional variables.

Example: Prove $\left[(A \rightarrow B) \wedge A\right] \rightarrow B$ is a tautology. The truth table is as follows:

A	B	$A \rightarrow B$	$(A \rightarrow B) \wedge A$	$[(A \rightarrow B) \wedge A] \rightarrow B$
True	True	True	True	True
True	False	False	False	True
False	True	True	False	True
False	False	True	False	True

Every value of $\left[(A \rightarrow B) \wedge A\right] \rightarrow B$ is "True", it is a tautology.

Contradiction

A Contradiction is a formula which is always false for every value of its propositional variables.

Example: Prove $(A \vee B) \wedge \left[(\neg A) \wedge (\neg B)\right]$ is a contradiction. The truth table is as follows:

A	B	$A \vee B$	$\neg A$	$\neg B$	$(\neg A) \wedge (\neg B)$	$(A \vee B) \wedge [(\neg A) \wedge (\neg B)]$
True	True	True	False	False	False	False
True	False	True	False	True	False	False
False	True	True	True	False	False	False
False	False	False	True	True	True	False

Every value of $(A \vee B) \wedge \left[(\neg A) \wedge (\neg B)\right]$ is "False", it is a contradiction.

Contingency

A Contingency is a formula which has both some true and some false values for every value of its propositional variables. Example: Prove $(A \vee B) \wedge (\neg A)$ a contingency. The truth table is as follows:

A	B	A ∨ B	¬ A	(A ∨ B) ∧ (¬ A)
True	True	True	False	False
True	False	True	False	False
False	True	True	True	True
False	False	False	True	False

Every value of $(A \vee B) \wedge (\neg A)$ has both "True" and "False", it is a contingency.

Propositional Equivalences

Two statements X and Y are logically equivalent if any of the following two conditions hold:

- The truth tables of each statement have the same truth values.
- The bi-conditional statement X⇔Y is a tautology.

Example: Prove $\neg(A \vee B)$ and $[(\neg A) \wedge (\neg B)]$ are equivalent.

Table: Testing by 1st method (Matching truth table).

A	B	A ∨ B	¬ (A ∨ B)	¬ A	¬ B	[(¬ A) ∧ (¬ B)]
True	True	True	False	False	False	False
True	False	True	False	False	True	False
False	True	True	False	True	False	False
False	False	False	True	True	True	True

The truth values of $\neg(A \vee B)$ and $[(\neg A) \wedge (\neg B)]$ are same, hence the statements are equivalent.

Table: Testing by 2nd method (Bi-conditionality).

A	B	¬ (A ∨ B)	[(¬ A) ∧ (¬ B)]	[¬ (A ∨ B)] ⇔ [(¬ A) ∧ (¬ B)]
True	True	False	False	True
True	False	False	False	True
False	True	False	False	True
False	False	True	True	True

As $[\neg(A \vee B)] \Leftrightarrow [(\neg A) \wedge (\neg B)]$ is a tautology, the statements are equivalent.

Inverse, Converse and Contra-positive

Implication / if-then ($\rightarrow$) is also called a conditional statement. It has two parts:

- Hypothesis, p.

- Conclusion, q.

It is denoted as $p \rightarrow q$.

Example of Conditional Statement: "If you do your homework, you will not be punished." Here, "you do your homework" is the hypothesis, p, and "you will not be punished" is the conclusion, q.

Inverse: An inverse of the conditional statement is the negation of both the hypothesis and the conclusion. If the statement is "If p, then q", the inverse will be "If not p, then not q". Thus the inverse of $p \rightarrow q$ is $\neg p \rightarrow \neg q$.

Example: The inverse of "If you do your homework, you will not be punished" is "If you do not do your homework, you will be punished."

Converse: The converse of the conditional statement is computed by interchanging the hypothesis and the conclusion. If the statement is "If p, then q", the converse will be "If q, then p". The converse of $p \rightarrow q$ is $q \rightarrow p$.

Example: The converse of "If you do your homework, you will not be punished" is "If you will not be punished, you do your homework".

Contra-positive: The contra-positive of the conditional is computed by interchanging the hypothesis and the conclusion of the inverse statement. If the statement is "If p, then q", the contra-positive will be "If not q, then not p". The contra-positive of $p \rightarrow q$ is $\neg q \rightarrow \neg p$.

Example: The Contra-positive of "If you do your homework, you will not be punished" is "If you are punished, you did not do your homework".

Duality Principle

Duality principle states that for any true statement, the dual statement obtained by interchanging unions into intersections (and vice versa) and interchanging Universal set into Null set (and vice versa) is also true. If dual of any statement is the statement itself, it is said self-dual statement.

Example: The dual of $(A \cap B) \cup C$ is $(A \cup B) \cap C$.

Normal Forms

We can convert any proposition in two normal forms: Conjunctive normal form and Disjunctive normal form.

Conjunctive Normal Form

A compound statement is in conjunctive normal form if it is obtained by operating AND among variables (negation of variables included) connected with ORs. In terms of set operations, it is a compound statement obtained by Intersection among variables connected with Unions.

Examples:

- $(A \vee B) \wedge (A \vee C) \wedge (B \vee C \vee D)$

- $(P \cup Q) \cap (Q \cup R)$

Disjunctive Normal Form

A compound statement is in disjunctive normal form if it is obtained by operating OR among variables (negation of variables included) connected with ANDs. In terms of set operations, it is a compound statement obtained by Union among variables connected with Intersections.

Examples:

- $(A \wedge B) \vee (A \wedge C) \vee (B \wedge C \wedge D)$
- $(P \cap Q) \cup (Q \cap R)$

Predicate Logic

Predicate Logic deals with predicates, which are propositions, consist of variables. A predicate is an expression of one or more variables determined on some specific domain. A predicate with variables can be made a proposition by either authorizing a value to the variable or by quantifying the variable. The following are some examples of predicates:

- Consider E(x, y) denote "x = y".
- Consider X(a, b, c) denote "a + b + c = 0".
- Consider M(x, y) denote "x is married to y."

Quantifier

The variable of predicates is quantified by quantifiers. There are two types of quantifier in predicate logic - Existential Quantifier and Universal Quantifier.

Existential Quantifier

If p(x) is a proposition over the universe U. Then it is denoted as $\exists x\ p(x)$ and read as "There exists at least one value in the universe of variable x such that p(x) is true. The quantifier $\exists$ is called the existential quantifier. There are several ways to write a proposition, with an existential quantifier, i.e., $(\exists x \in A)p(x)$ or $\exists x \in A$ such that p(x) or $(\exists x)p(x)$ or p(x) is true for some $x \in A$.

Universal Quantifier

If p(x) is a proposition over the universe U. Then it is denoted as $\forall x, p(x)$ and read as "For every $x \in U, p(x)$ is true." The quantifier $\forall$ is called the Universal Quantifier. There are several ways to write a proposition, with a universal quantifier. $\forall x \in A, p(x)$ or p(x), $\forall x \in A$ Or $\forall x, p(x)$ or p(x) is true for all $x \in A$.

Negation of Quantified Propositions

When we negate a quantified proposition, i.e., when a universally quantified proposition is negated, we obtain an existentially quantified proposition, and when an existentially quantified proposition is negated, we obtain a universally quantified proposition. The two rules for negation of quantified proposition are as follows. These are also called DeMorgan's Law.

Example: Negate each of the following propositions:

1. $\forall x\, p(x) \wedge \exists y\, q(y)$

$$\sim .\forall x\, p(x) \wedge \exists y\, q(y))$$

Solution: $\cong \sim \forall x\, p(x) \vee \sim \exists y q\,(y) \quad (\therefore \sim (p \wedge q) = \sim p \vee \sim q)$

$$\cong \exists x\, \sim p(x) \vee \forall y \sim q(y)$$

2. $(\exists x \in U)\,(x+6=25)$

$$\sim (\exists x \in U)\,(x+6=25)$$

Solution: $\cong \forall x \in U \sim\, (x+6)=25$

$$\cong (\forall x \in U)\,(x+6) \neq 25$$

3. $\sim (\exists x\, p(x) \vee \forall y\, q(y)$

$$\sim (\exists x\, p(x) \vee \forall y\, q(y))$$

Solution: $\cong \sim \exists x\, p(x) \wedge \sim \forall y\, q(y)\; (\therefore \sim (p \vee q) = \sim p \wedge \sim q)$

$$\cong \forall x \sim\, p(x) \wedge \exists y \sim q(y))$$

Propositions with Multiple Quantifiers

The proposition having more than one variable can be quantified with multiple quantifiers. The multiple universal quantifiers can be arranged in any order without altering the meaning of the resulting proposition. Also, the multiple existential quantifiers can be arranged in any order without altering the meaning of the proposition.

The proposition which contains both universal and existential quantifiers, the order of those quantifiers can't be exchanged without altering the meaning of the proposition, e.g., the proposition $\exists x\, \forall y\, p(x,y)$ means "There exists some x such that p (x, y) is true for every y."

Example: Write the negation for each of the following. Determine whether the resulting statement is true or false. Assume U = R.

1. $\forall x \exists m(x^2 < m)$

Solution: Negation of $\forall x \exists m(x^2<m)$ is $\exists x \forall m\,(x^2 \geq m)$. The meaning of $\exists x \forall m\,(x^2 \geq m)$ is that there exists for some x such that $x^2 \geq m$, for every m. The statement is true as there is some greater x such that $x^2 \geq m$, for every m.

2. $\exists m \forall x \left(x^2 < m\right)$

Solution: Negation of $\exists m \forall x (x^2<m)$ is $\forall m \exists x (x^2 \geq m)$. The meaning of $\forall m \exists x (x^2 \geq m)$ is that for every m, there exists for some x such that $x^2 \geq m$. The statement is true as for every m, there exists for some greater x such that $x^2 \geq m$.

References

- Higher-math-online, mathematics: whitman.edu, Retrieved 29, June 2020
- Sec-logic-proofs: openmathbooks.org, Retrieved 12, April 2020
- Discrete-mathematics-probability: tutorialspoint.com, Retrieved 06, January 2020
- Discrete-mathematics-counting-theory: tutorialspoint.com, Retrieved 14, August 2020
- Predicate-logic: javatpoint.com, Retrieved 05, May 2020

2

Understanding Graph Theory

The branch of mathematics which focuses on the networks of points connected by lines is known as graph theory. The different types of graphs studied within this field are null graph, simple graph, trivial graph and undirected graph. All these diverse types of graphs have been carefully analyzed in this chapter.

A graph is a diagram of points and lines connected to the points. It has at least one line joining a set of two vertices with no vertex connecting itself. The concept of graphs in graph theory stands up on some basic terms such as point, line, vertex, edge, degree of vertices, properties of graphs, etc. Here, in this chapter, we will cover these fundamentals of graph theory.

Point

A point is a particular position in a one-dimensional, two-dimensional, or three-dimensional space. For better understanding, a point can be denoted by an alphabet. It can be represented with a dot. For Example:

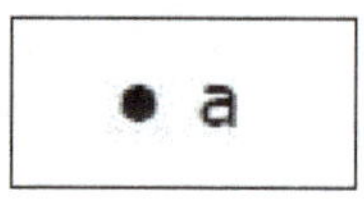

Here, the dot is a point named 'a'.

Line

A Line is a connection between two points. It can be represented with a solid line. For Example:

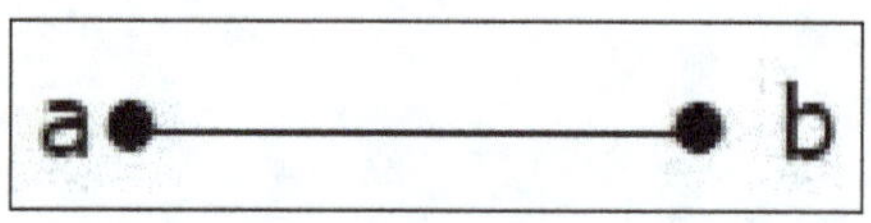

Here, 'a' and 'b' are the points. The link between these two points is called a line.

Vertex

A vertex is a point where multiple lines meet. It is also called a node. Similar to points, a vertex is also denoted by an alphabet. For Example:

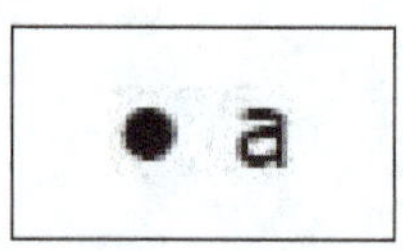

Here, the vertex is named with an alphabet 'a'.

Edge

An edge is the mathematical term for a line that connects two vertices. Many edges can be formed from a single vertex. Without a vertex, an edge cannot be formed. There must be a starting vertex and an ending vertex for an edge. For Example:

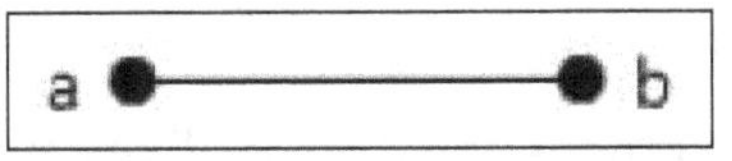

Here, 'a' and 'b' are the two vertices and the link between them is called an edge.

Graph

A graph 'G' is defined as G = (V, E) Where V is a set of all vertices and E is a set of all edges in the graph. For Example:

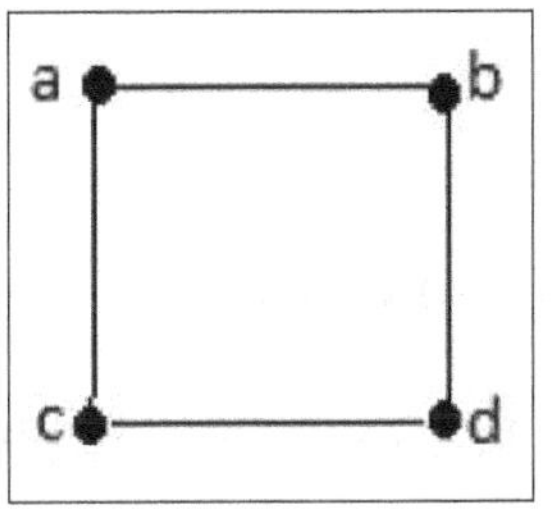

In the above example, ab, ac, cd, and bd are the edges of the graph. Similarly, a, b, c, and d are the vertices of the graph.

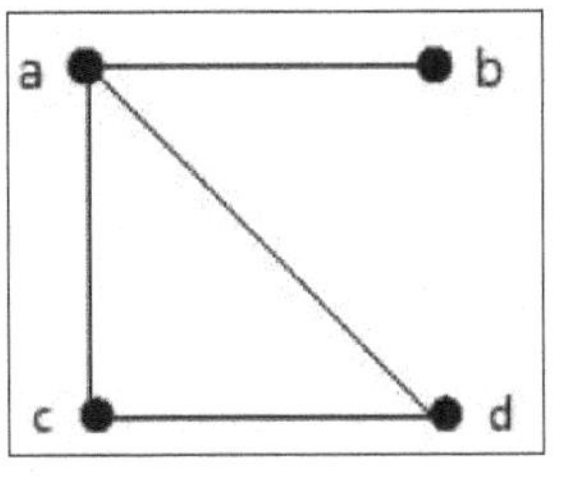

In this graph, there are four vertices a, b, c, and d, and four edges ab, ac, ad, and cd.

Loop

In a graph, if an edge is drawn from vertex to itself, it is called a loop. For Example:

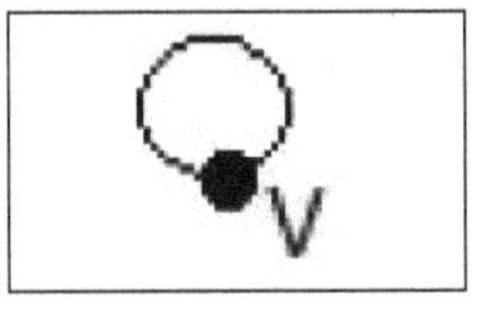

V is a vertex for which it has an edge (V, V) forming a loop.

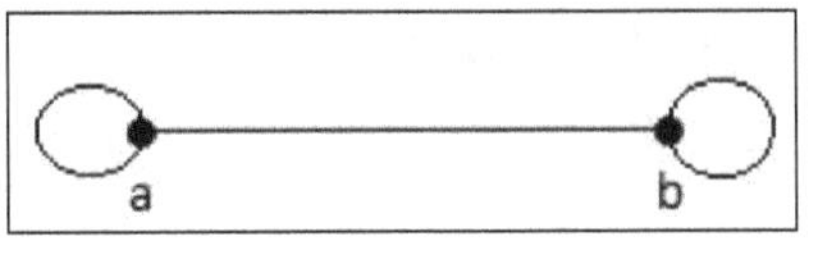

In this graph, there are two loops which are formed at vertex a, and vertex b.

Degree of Vertex

It is the number of vertices adjacent to a vertex V.

Notation – deg(V).

In a simple graph with n number of vertices, the degree of any vertices is – $deg(v) \leq n - 1 \ \forall \ v \in G$.

A vertex can form an edge with all other vertices except by itself. So the degree of a vertex will be up to the number of vertices in the graph minus 1. This 1 is for the self-vertex as it cannot form a loop by itself. If there is a loop at any of the vertices, then it is not a Simple Graph.

Degree of vertex can be considered under two cases of graphs:

- Undirected Graph
- Directed Graph

Degree of Vertex in an Undirected Graph

An undirected graph has no directed edges. Consider the following examples. Take a look at the following graph:

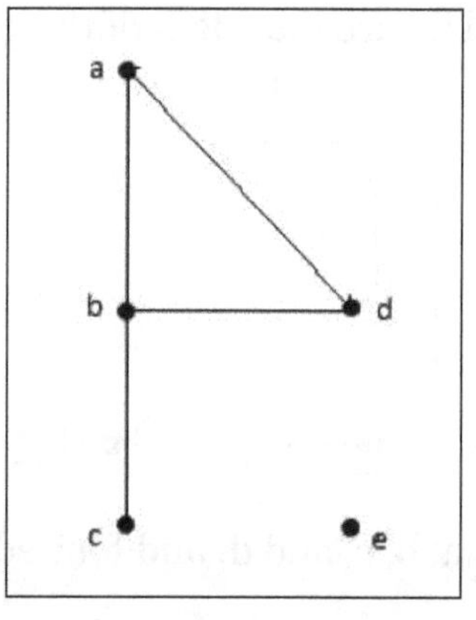

In the above Undirected Graph,

- deg(a) = 2, as there are 2 edges meeting at vertex 'a'.
- deg(b) = 3, as there are 3 edges meeting at vertex 'b'.
- deg(c) = 1, as there is 1 edge formed at vertex 'c'
- So 'c' is a pendent vertex.
- deg(d) = 2, as there are 2 edges meeting at vertex 'd'.
- deg(e) = 0, as there are 0 edges formed at vertex 'e'.
- So 'e' is an isolated vertex.

Example: Take a look at the following graph:

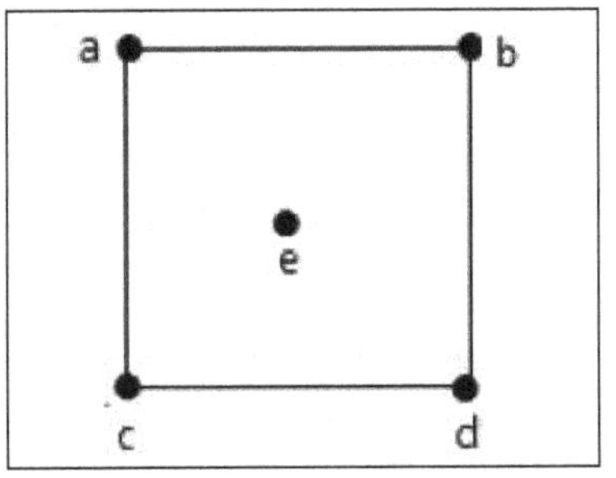

deg(a) = 2, deg(b) = 2, deg(c) = 2, deg(d) = 2, and deg(e) = 0.

The vertex 'e' is an isolated vertex. The graph does not have any pendent vertex.

Degree of Vertex in a Directed Graph

In a directed graph, each vertex has an indegree and an outdegree.

Indegree of a Graph

- Indegree of vertex V is the number of edges which are coming into the vertex V.
- Notation – deg−(V).

Outdegree of a Graph

- Outdegree of vertex V is the number of edges which are going out from the vertex V.
- Notation – deg+(V).

Consider the following examples.

Take a look at the following directed graph. Vertex 'a' has two edges, 'ad' and 'ab', which are going outwards. Hence its outdegree is 2. Similarly, there is an edge 'ga', coming towards vertex 'a'. Hence the indegree of 'a' is 1.

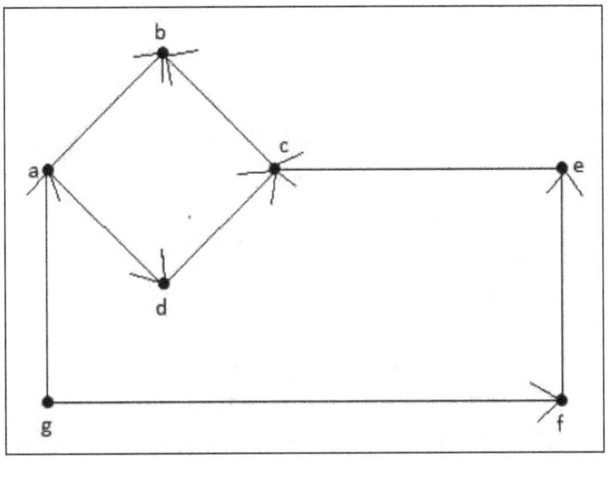

The indegree and outdegree of other vertices are shown in the following table:

Vertex	Indegree	Outdegree
a	1	2
b	2	0
c	2	1
d	1	1
e	1	1
f	1	1
g	0	2

Example: Take a look at the following directed graph. Vertex 'a' has an edge 'ae' going outwards from vertex 'a'. Hence its outdegree is 1. Similarly, the graph has an edge 'ba' coming towards vertex 'a'. Hence the indegree of 'a' is 1.

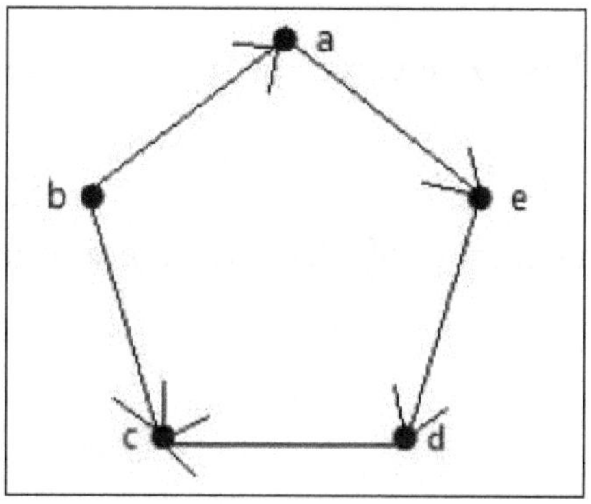

The indegree and outdegree of other vertices are shown in the following table:

Vertex	Indegree	Outdegree
a	1	1
b	0	2
c	2	0
d	1	1
e	1	1

Pendent Vertex

By using degree of a vertex, we have two special types of vertices. A vertex with degree one is called a pendent vertex.

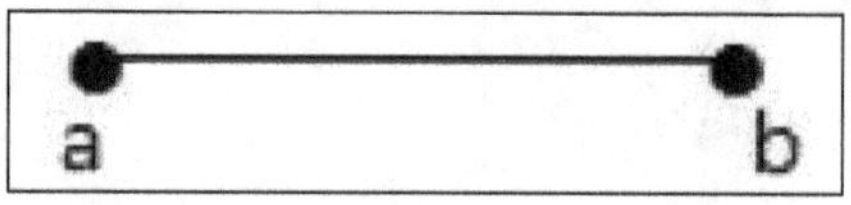

Here, vertex 'a' and vertex 'b' have a connected edge 'ab'. So with respect to the vertex 'a', there is only one edge towards vertex 'b' and similarly with respect to the vertex 'b', there is only one edge towards vertex 'a'. Finally, vertex 'a' and vertex 'b' has degree as one which are also called as the pendent vertex.

Isolated Vertex

A vertex with degree zero is called an isolated vertex. For Example:

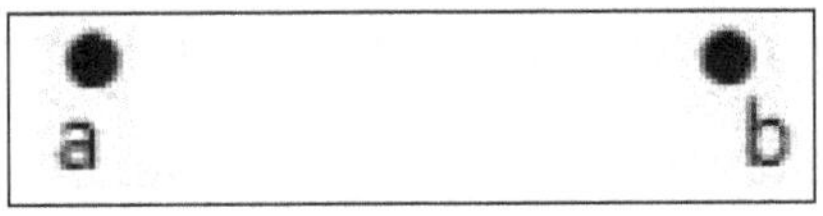

Here, the vertex 'a' and vertex 'b' has a no connectivity between each other and also to any other vertices. So the degree of both the vertices 'a' and 'b' are zero. These are also called as isolated vertices.

Adjacency

Here are the norms of adjacency:

- In a graph, two vertices are said to be adjacent, if there is an edge between the two vertices. Here, the adjacency of vertices is maintained by the single edge that is connecting those two vertices.
- In a graph, two edges are said to be adjacent, if there is a common vertex between the two edges. Here, the adjacency of edges is maintained by the single vertex that is connecting two edges. Examples:

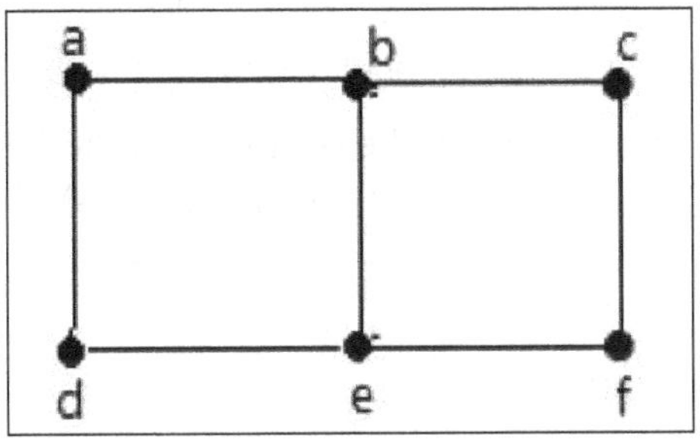

- 'a' and 'b' are the adjacent vertices, as there is a common edge 'ab' between them.
- 'a' and 'd' are the adjacent vertices, as there is a common edge 'ad' between them.
- ab' and 'be' are the adjacent edges, as there is a common vertex 'b' between them.
- be' and 'de' are the adjacent edges, as there is a common vertex 'e' between them.

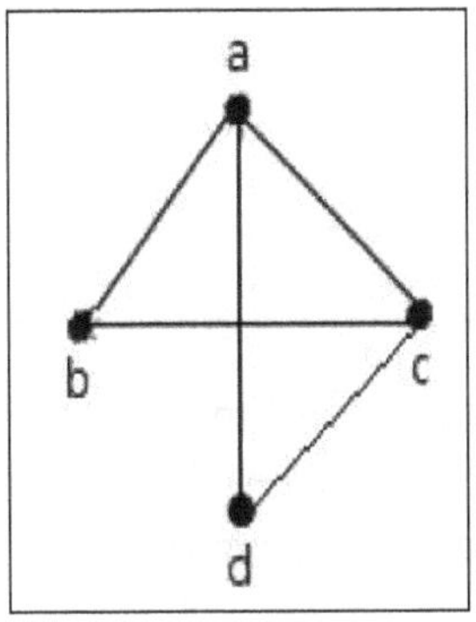

- 'a' and 'd' are the adjacent vertices, as there is a common edge 'ad' between them.
- 'c' and 'b' are the adjacent vertices, as there is a common edge 'cb' between them.
- 'ad' and 'cd' are the adjacent edges, as there is a common vertex 'd' between them.
- 'ac' and 'cd' are the adjacent edges, as there is a common vertex 'c' between them.

Parallel Edges

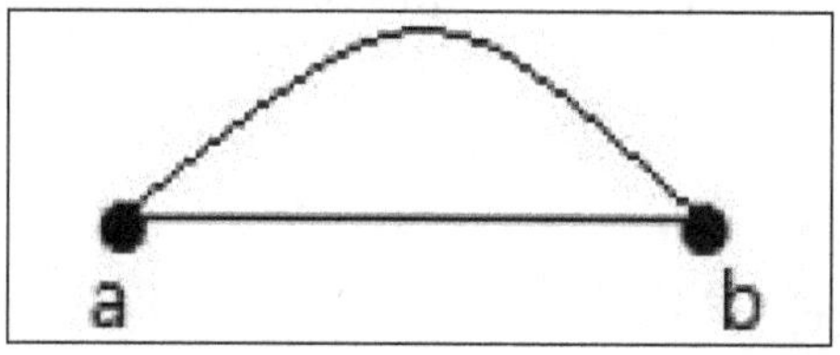

In a graph, if a pair of vertices is connected by more than one edge, then those edges are called parallel edges.

'a' and 'b' are the two vertices which are connected by two edges 'ab' and 'ab' between them. So it is called as a parallel edge.

Graph theory is an important branch of discrete mathematics, which has a number of applications in diverse areas such as computer science, natural sciences, social sciences, engineering, ecological and management sciences.

- It is believed that graph theory had its beginning in 1736 when Euler considered the (general case of the) Konigsberg bridge problem which is stated below.

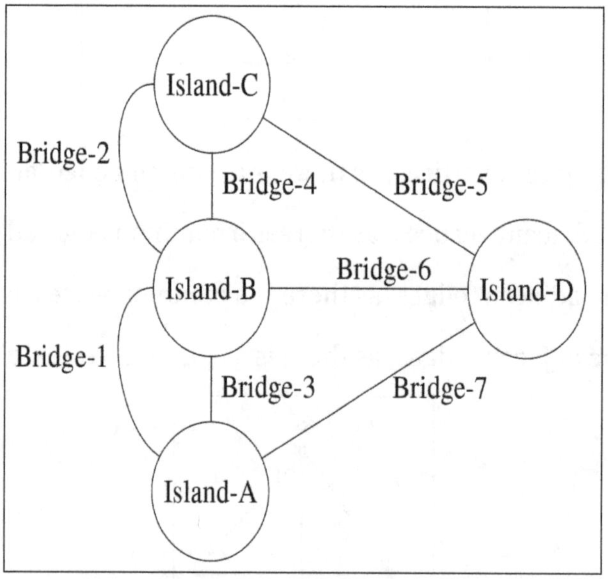

Consider the picture given above. Is there a walking route starting at an Island-X (say), that crosses each of the seven bridges (of Königsberg) exactly once and returns back to Island-X? It took 200 years before the first book on graph theory was written. This was done by König in 1936.

- Since then graph theory has developed into an extensive and popular branch of mathematics, which has been applied to many problems in mathematics, computer science, and other scientific and not-so-scientific areas.
    - A directed graph (in short, digraph) is an ordered pair X = (V,E), where V is a non-empty set, called the vertex set, and E is a subset of V ×V , called the edge set. Note that we also use V(X) and E(X) to denote vertex set and edge set of digraph X, respectively.
    - If e = (u, v) ∈ E with u 6= v then, the edge e is said to be incident from u to v or u is said to be the initial vertex and v the terminal vertex of e. An edge e = (u,u) is called a loop.

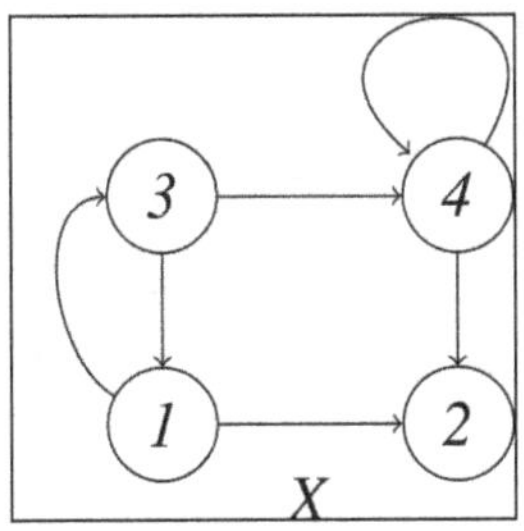

The above graph X is a directed graph with vertex set V = {1,2,3,4} and edge set,

E = {(1,2),(1,3),(3,1),(3,4),(4,2),(4,4)}.

For the edge (1,2), 1 is the initial vertex and 2 is the terminal vertex. There is a self-loop a the vertex 4.

- A digraph is called a graph if (u, v) ∈ E whenever (v,u) ∈ E, for any two elements u, v ∈ V. If u, v ∈ V, then an edge between u and v in the graph X is denoted by e = {u, v} and in this case, we say that e is incident with u and v or the vertex u is adjacent to the vertex v, Or vice-versa.
- If two edges e1 and e2 are incident with a common vertex then the two edges are said to be adjacent.
- For any finite set S, let |S| denote the number of elements in S. Then a graph/digraph is said to be finite, if |V| (called the order of X) and |E| (called the size of X) are finite.
- There is a possibility of having more than one edge between two given vertices of a graph, called multiple edges. If a graph contains finite number of vertices but an infinite number of edges then, the graph contains at least two vertices which have infinite number of edges between them. The graph below contains multiple edges between the vertices 1 and 2 as well as between 2 and 3.

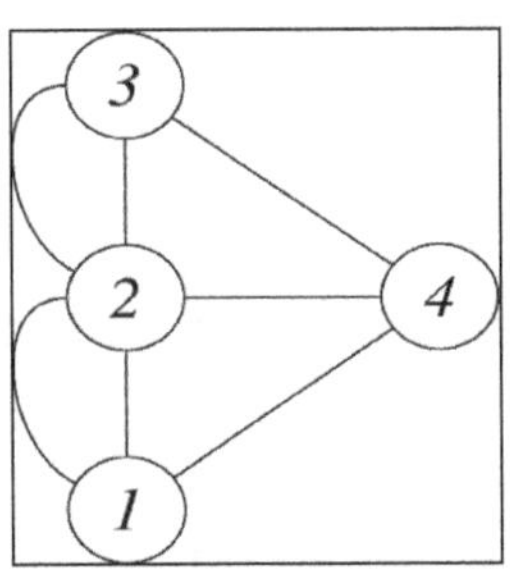

- A graph is called simple if it has no loops and no multiple edges.
- Let X = (V,E) be a graph. Then the degree of a vertex v ∈ V, denoted d(v), is the number of edges incident with it. In case v is a vertex of a digraph X, one defines in-degree of v, denoted d+(v), as the number of edges that have v as a terminal vertex and out-degree of v, denoted d−(v), as the number of edges that have v as an initial vertex.

For Example: 1. Let A = {1,2,3,4} and B, the set of all two element subsets of A. Now, we construct a graph with B as the vertex set and there is an edge between two vertices whenever their intersection is non-empty.

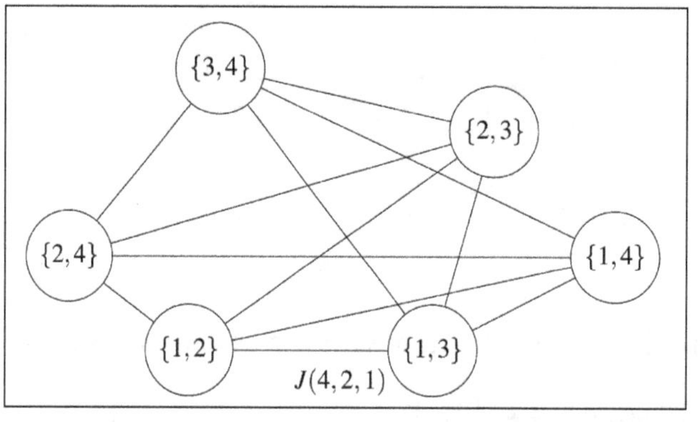

In general, let n, k,i ∈ $\mathbb{Z}^+$ and n ≥ k ≥ i. Let S be a fixed set of cardinality n. Then, the Johnson graph, denoted J(n, k,i), is defined as follows:

- The vertices are the k-element subsets of S.
- Two vertices are adjacent, whenever their intersection has size i.

2. Let A = {1,2,3,4,5} and B, the set of all two element subsets of A. Now, we construct a graph with B as the vertex set and there is an edge between two vertices whenever their intersection is empty. This graph is commonly known as the Petersen graph.

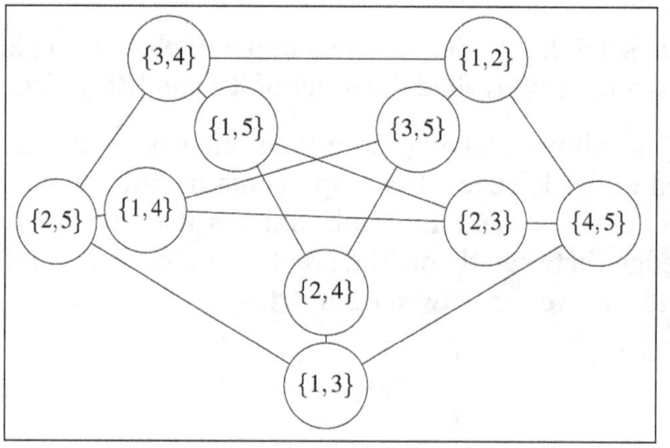

3. Let k ∈ ℕ. Then, the k-cube graph, denoted Q_k, is defined as follows:

- The vertex set of Q_k is the set of all 2^k binary k-tuples.
- Two vertices are adjacent, whenever they differ in exactly in one coordinate position.

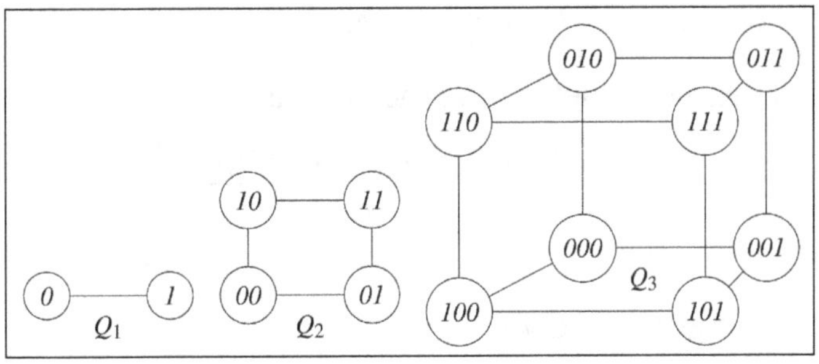

Theorem: (First theorem on graph theory (Hand shaking lemma)). The sum of the degrees of all vertices is equal to twice the number of edges.

Proof:

- We can prove the result by induction on number of edges.
- If there are no edges in the graph, then the degree of each vertex is zero, thus result follows trivially.
- Assume the result is true for all graphs with k edges.
- Let X be a graph with k +1 edges and Y be a graph obtained from X by removing an edge. Then, the graph Y has the same vertex set as that of X but contains k edges. Thus, by induction hypothesis, the number of edges in Y is even. Hence, by adding an edge back to Y, say e = {u, v}, to get the graph X, we see that the degree of the incident vertices u and v (the in-degree and the out-degree in case of directed graph) increases by 1 and hence their contribution to the sum of the degrees increases by 2. Thus, the required result follows:

  The sum of the in degrees of all vertices = the sum of out degrees of all vertices = the number of edges

Example: A knight is a chess piece that can move either.

- Two spaces horizontally and one space vertically.
- One space horizontally and two spaces vertically.

Consider a n × n chess board. We define a knight graph, denoted by knight(n), as the graph which has $V=\{(i,j)\mid 1\le i,j\le n\}$ as its vertex set and $E=\{(i,j),(k,l)\mid \{|i-k|,|j-l|\}=\{1,2\}\}$ as its edge set. That is, there is a vertex for every square of the board and there is an edge between two vertices exactly when there is a knight's move from one to the other.

For n ≥ 8, the knight graph knight(n) has:

- 4 vertices of degree 2 $(\{(1,1),(1,n),(n,1),(n,n)\})$.
- 8 vertices of degree 3 $(\{(1,2),(2,1),(1,n-1),(n-1,1),(2,n),(n,2),(n-1,n),(n,n-1)\})$.
- 4(n−3) vertices of degree 4 $(\{(1,i),(i,1) : 3 \le i \le n-2\} \cup \{(2,2),(2,n-1),(n-1,2),$ $(n-1,n-1)\})$.

- $4(n-4)$ vertices of degree 6 - determine the vertices.
- $(n-4)^2$ vertices of degree 8 - determine the vertices.
- The number of edges = $\dfrac{4.2+8.3+4(n-3).4+4(n-6).6+(n-4)^2.8}{2} = 4n^2-12n+8$

$$=4(n-1)(n-2).$$

Definition: Let X = (V,E) be a graph and $v \in V$

- If deg(v) = 0 then v is called an isolated vertex.
- If deg(v) = 1 then v is called a pendant vertex (Pendent vertex is adjacent to exactly one vertex).
- If deg(v) is even then v is called an even vertex.
- If deg(v) is odd then v is called an odd vertex.

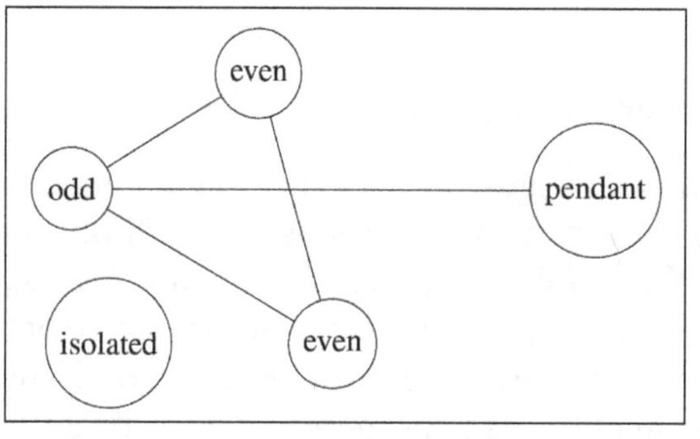

- If X be a graph on n vertices. Then the sequence $\{d_1, d_2, ..., d_n\}$ with $d_1 \le d_2 \le \cdots \le d_n$ is called the degree sequence of X, where $d_1, d_2, ..., d_n$ are the degrees of the vertices of X. The degree sequence of the above graph is (0,1,2,2,3).
- Let X be a simple undirected finite graph. We use the following notations:

  $\Delta(X) = \max\{\deg(v) \mid v \in V\}$

  $\delta(X) = \min\{\deg(v) \mid v \in V\}$

Let X be a simple undirected finite graph on n vertices. Then using theorem, we have

$$0 \le \delta(X) \le \deg(v) \le \Delta(X) \le n-1, \text{ and } \delta(X)|V| \le 2|E|, \Delta(X)|V| \ge 2|E|.$$

Corollary: The number of odd vertices in a graph is always even.

Proof: Let X = (V,E) be a graph. Let d(v) denote the degree of $v \in V$. From Theorem, we have,

$$\underbrace{\sum_{\text{even vertices}} d(v_i)}_{A} + \underbrace{\sum_{\text{odd vertices}} d(v_j)}_{B} = 2|E|$$

Hence,

$$\sum_{\text{odd vertices}} d(v_j) = \text{an even number.}$$

From now onwards, unless specified otherwise, all the graphs which we consider are assumed to be finite and simple.

Types of Graphs

Though, there are a lot of different types of graphs depending upon the number of vertices, number of edges, interconnectivity, and their overall structure, some of such common types of graphs are as follows:

1. Null Graph: A null graph is a graph in which there are no edges between its vertices. A null graph is also called empty graph. For example:

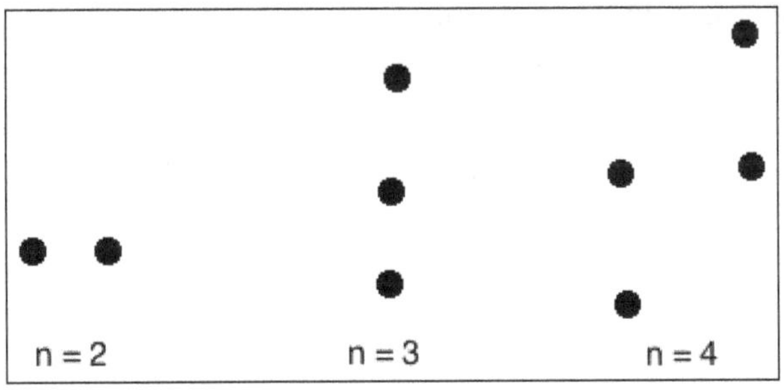

A null graph with n vertices is denoted by Nn.

2. Trivial Graph: A trivial graph is the graph which has only one vertex. In the above graph, there is only one vertex 'v' without any edge. Therefore, it is a trivial graph.

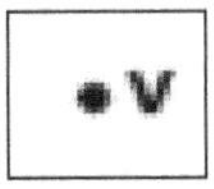

3. Simple Graph: A simple graph is the undirected graph with no parallel edges and no loops. A simple graph which has n vertices, the degree of every vertex is at most n -1.

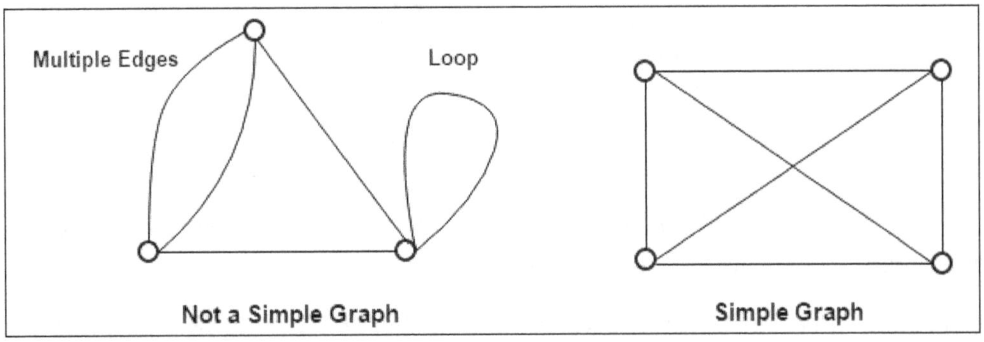

In the above example, First graph is not a simple graph because it has two edges between the vertices A and B and it also has a loop. Second graph is a simple graph because it does not contain any loop and parallel edges.

4. Undirected Graph: An undirected graph is a graph whose edges are not directed. For example:

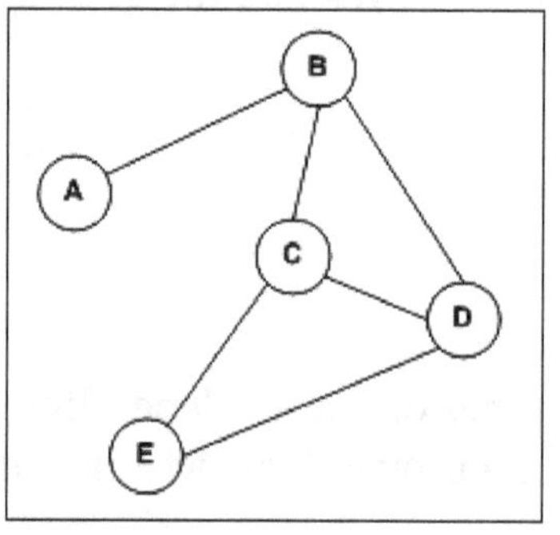

Since there are no directed edges, therefore it is an undirected graph.

5. Directed Graph: A directed graph is a graph in which the edges are directed by arrows. Directed graph is also known as digraphs. For example:

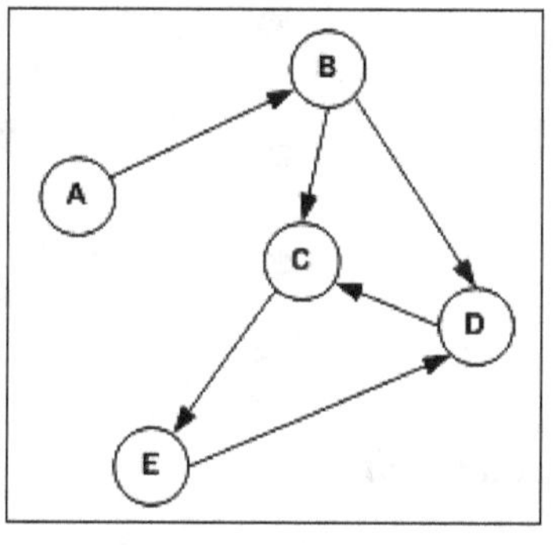

In the above graph, each edge is directed by the arrow. A directed edge has an arrow from A to B, means A is related to B, but B is not related to A.

6. Complete Graph: A graph in which every pair of vertices is joined by exactly one edge is called complete graph. It contains all possible edges. A complete graph with n vertices contains exactly nC2 edges and is represented by Kn.

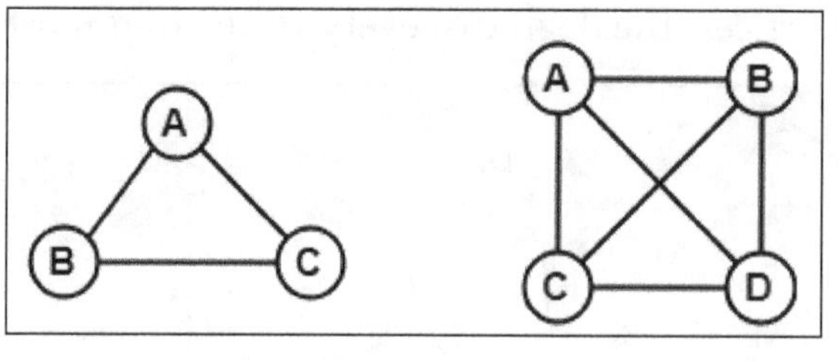

Since each vertex in the graph is connected with all the remaining vertices through exactly one edge therefore, both graphs are complete graph.

7. Connected Graph: A connected graph is a graph in which we can visit from any one vertex to any other vertex. In a connected graph, at least one edge or path exists between every pair of vertices.

For example:

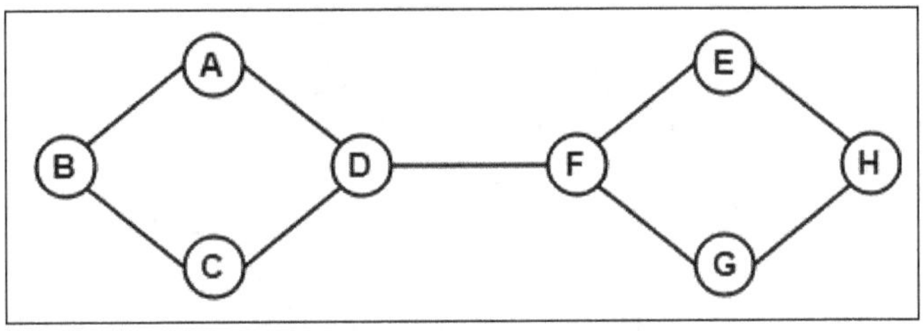

We can traverse from any one vertex to any other vertex. It means there exists at least one path between every pair of vertices therefore, it a connected graph.

8. Disconnected Graph: A disconnected graph is a graph in which any path does not exist between every pair of vertices. For example:

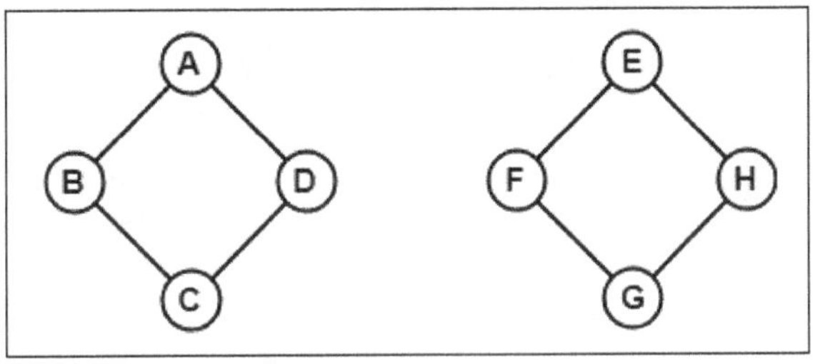

The graph consists of two independent components which are disconnected. Since it is not possible to visit from the vertices of one component to the vertices of other components therefore, it is a disconnected graph.

9. Regular Graph: A Regular graph is a graph in which degree of all the vertices is same. If the degree of all the vertices is k, then it is called k-regular graph.

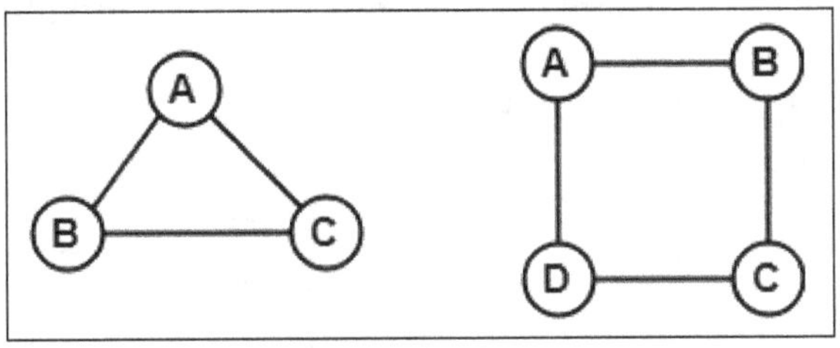

All the vertices have degree 2. Therefore they are called 2- Regular graph.

10. Cyclic Graph: A graph with 'n' vertices (where, n>=3) and 'n' edges forming a cycle of 'n' with all its edges is known as cycle graph. A graph containing at least one cycle in it is known as a cyclic graph. In the cycle graph, degree of each vertex is 2. The cycle graph which has n vertices is denoted by Cn.

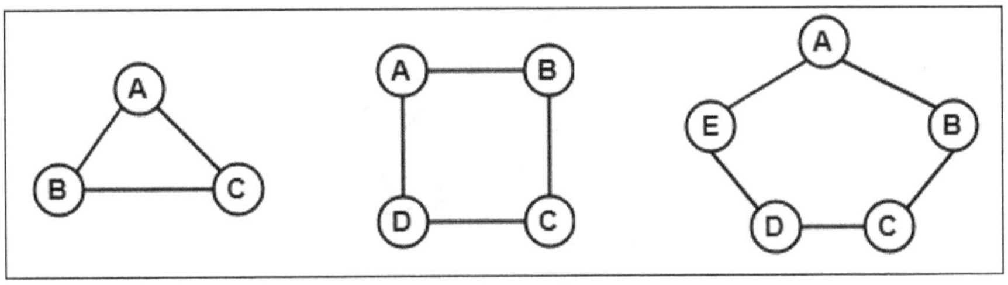

All the vertices have degree 2. Therefore they all are cyclic graphs.

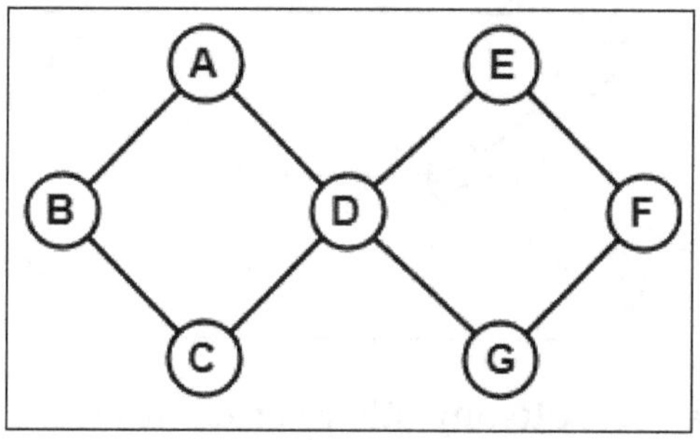

The above graph contains two cycles in it therefore, it is a cyclic graph.

11. Acyclic Graph: A graph which does not contain any cycle in it is called as an acyclic graph.

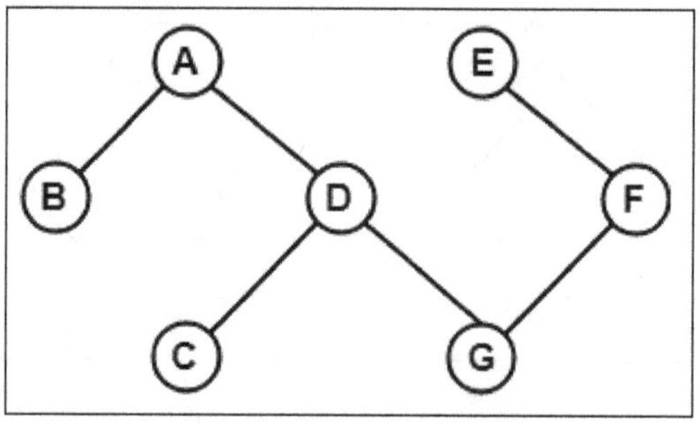

The above graph does not contain any cycle in it therefore, it is an acyclic graph.

12. Bipartite Graph: A bipartite graph is a graph in which the vertex set can be partitioned into two sets such that edges only go between sets, not within them. A graph G (V, E) is called bipartite graph if its vertex-set V(G) can be decomposed into two non-empty disjoint subsets V1(G) and V2(G) in such a way that each edge e ∈ E(G) has its one last joint in V1(G) and other last point in V2(G).

The partition V = V1 ∪ V2 is known as bipartition of G.

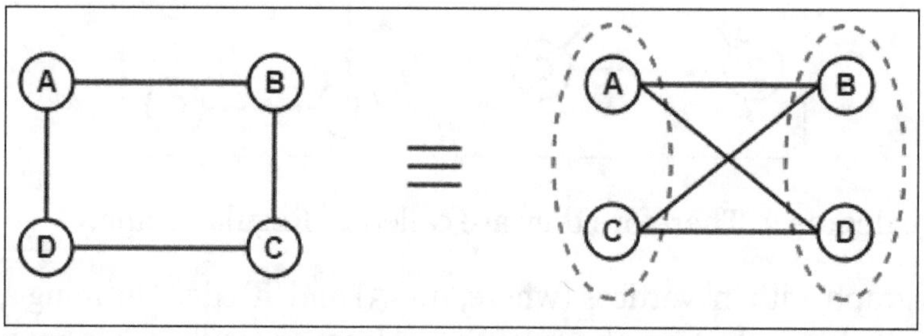

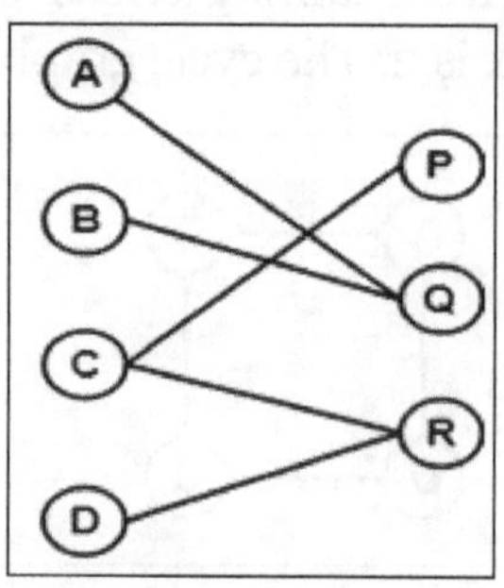

13. Complete Bipartite Graph: A complete bipartite graph is a bipartite graph in which each vertex in the first set is joined to each vertex in the second set by exactly one edge. A complete bipartite graph is a bipartite graph which is complete.

Complete Bipartite graph = Bipartite graph + Complete graph

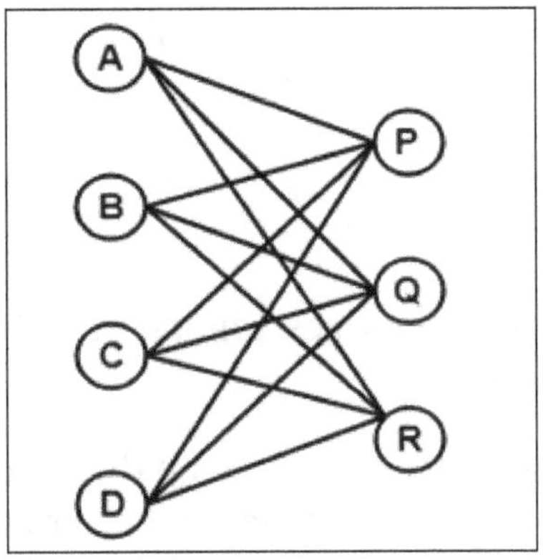

The above graph is known as $K_{4,3}$.

14. Star Graph: A star graph is a complete bipartite graph in which n-1 vertices have degree 1 and a single vertex have degree (n -1). This exactly looks like a star where (n - 1) vertices are connected to a single central vertex. A star graph with n vertices is denoted by S_n.

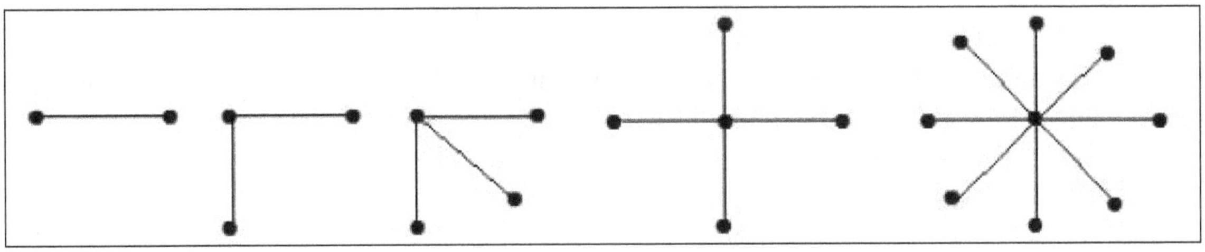

Out of n vertices, all the (n-1) vertices are connected to a single vertex. Hence, it is a star graph.

15. Weighted Graph: A weighted graph is a graph whose edges have been labeled with some weights or numbers.The length of a path in a weighted graph is the sum of the weights of all the edges in the path.

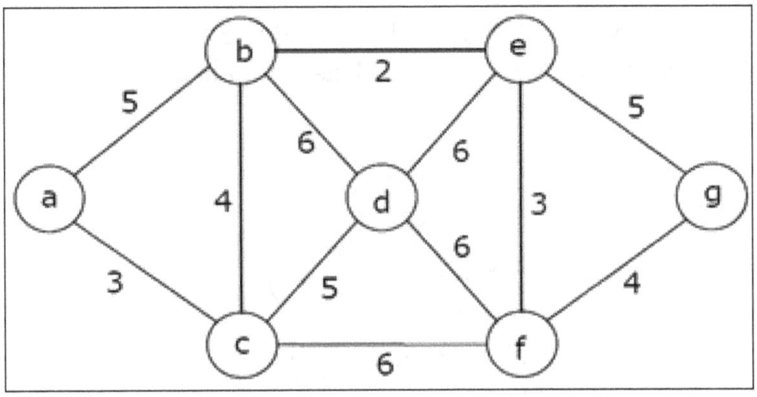

If path is a -> b -> c -> d -> e -> g then the length of the path is 5 + 4 + 5 + 6 + 5 = 25.

16. Multi-graph: A graph in which there are multiple edges between any pair of vertices or there are edges from a vertex to itself (loop) is called a multi - graph. For Example:

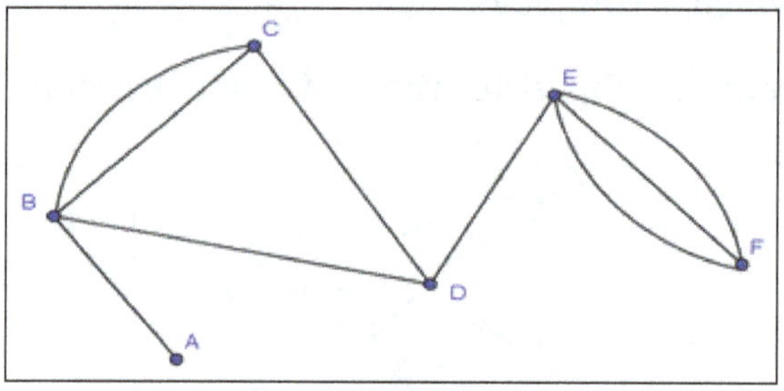

Vertex-set B and C are connected with two edges. Similarly, vertex sets E and F are connected with 3 edges. Therefore, it is a multi-graph.

17. Planar Graph: A planar graph is a graph that we can draw in a plane in such a way that no two edges of it cross each other except at a vertex to which they are incident. For example:

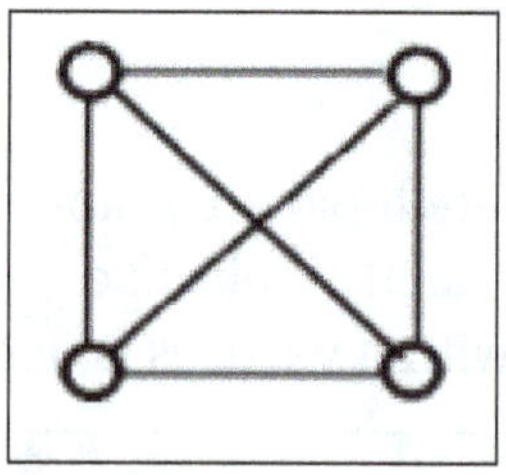

The above graph may not seem to be planar because it has edges crossing each other. But we can redraw the above graph. The three plane drawings of the above graph are:

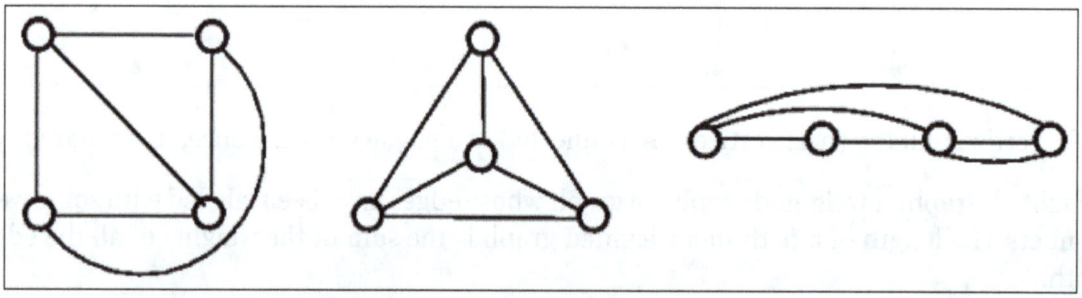

18. Non - Planar Graph: A graph that is not a planar graph is called a non-planar graph. In other words, a graph that cannot be drawn without at least on pair of its crossing edges is known as non-planar graph.

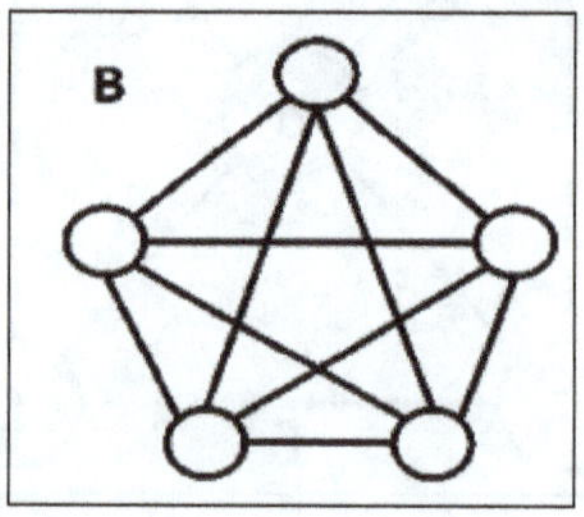

The above graph is a non - planar graph.

Connectedness of Graphs

Connected Graph X is said to connected if any two vertices of X are connected. Note that Complete graphs, complete bipartite graphs, wheel graphs, cyclic graphs, path graphs and Johnson graphs etc. are a few examples of connected graphs. A graph that is not connected is the union of two or more connected subgraphs, each pair of which has no vertex in common. These disjoint connected subgraphs are called the connected components of the graph. The number of components of a graph X is denoted by C(X). If X is connected then C(X) = 1. Let e be an edge of a graph X then it can be easily observed that $C(X) \leq C(X \setminus \{e\}) \leq C(X) + 1$. A direct application of the definition of a connected/disconnected graph gives the following result and hence the proof is omitted.

Theorem: A graph is disconnected if and only if its vertex set V can be partitioned into two non-empty disjoint subsets V_1 and V_2 such that there exists no edge in X whose one end vertex is in V_1 and other is in V_2.

Theorem: If a graph (connected or disconnected) has exactly two vertices of odd degree then there must be a path joining these vertices.

Proof: Let X be a graph that has exactly two vertices, say u and v of odd degree. Let C be a connected component of X containing the vertex u. Since, the number of vertices of odd degree in any graph is even, C has to contain another vertex of odd degree. But, then v is the only other vertex in X of odd degree and hence v lies in the component C. Thus, there is a path from u to v. Before proving the next theorem, we need the following inequality.

Lemma: Let $n_1, n_2, \ldots, n_k \in \mathbb{N}$. Then,

$$\sum_{i=1}^{k} n_i^2 \leq \left(\sum_{i=1}^{k} n_i\right)^2 - (k-1)\left(2\sum_{i=1}^{k} n_i - k\right)$$

Proof: We have,

$$(n_1 - 1) + (n_2 - 1) + \ldots + (n_k - 1) = (n_1 + n_2 + \cdots + n_k) - k$$

Squaring on both sides, we get

$$\left[(n_1 - 1) + (n_2 - 1) + \ldots + (n_k - 1)\right]^2 = \left[(n_1 + n_2 + \ldots + n_k) - k\right]^2$$

$$\sum_{i=1}^{k} (n_1 - 1)^2 + \sum_{i=1}^{k} \sum_{j=1, j\neq i}^{k} (n_i - 1)(n_j - 1) = \left(\sum_{i=1}^{k} n_1\right)^2 - 2k\left(\sum_{i=1}^{k} n_1\right) + k^2$$

$$\sum_{i=1}^{k} n_i^2 - 2\left(\sum_{i=1}^{k} n_i\right) + k + \underbrace{\sum_{i=1}^{k} \sum_{j=1, j\neq i}^{k} (n_i - 1)(n_j - 1)}_{\text{positive}} = \left(\sum_{i=1}^{k} n_i\right)^2 - 2k\left(\sum_{i=1}^{k} n_i\right) + k^2$$

Hence, we have

$$\sum_{i=1}^{k} n_i^2 \le 2\left(\sum_{i=1}^{k} n_i\right) - k + \left(\sum_{i=1}^{k} n_i\right)^2 - 2k\left(\sum_{i=1}^{k} n_i\right) + k^2$$

Or equivalently,

$$\sum_{i=1}^{k} n_i^2 \le \left(\sum_{i=1}^{k} n_i\right) - (k-1)\left(2\sum_{i=1}^{k} n_i - k\right)$$

Theorem: A simple graph with n vertices and k components can have at most have (n−k)(n−k +1)/2 edges.

Proof: Let X be a graph with k components. Let n_i be the number of vertices in the i th component, where $1 \le i \le k$. Then, the number of edges in the graph is equal to sum of the edges in each of its components. Thus, X has maximum number of edges if each component is a complete graph.

Hence, the maximum possible number of edges in the graph X is $\sum_{i=1}^{k} \frac{n_1(n_1-1)}{2}$. But from Lemma we have,

$$\frac{1}{2}\sum_{i=1}^{k} n_i^2 - \frac{1}{2}\sum_{i=1}^{k} n_i \le \frac{1}{2}\left[\left(\sum_{i=1}^{k} n_i\right)^2 - (k-1)\left(2\sum_{i=1}^{k} n_i - k\right)\right] - \frac{1}{2}\sum_{i=1}^{k} n_i$$

$$= \frac{1}{2}(n-k)(n-k+1)$$

Theorem: Let X be a simple graph with diameter d(X). If d(X) ≥ 3 then show that d(X c) is ≤ 3.

Proof: Since d(X) ≥ 3, there exist two non-adjacent vertices, say u, v in X, such that u and v have no common neighbor. Then, for each x ∈ V(X) \ {u, v}, either u or v or both is a non-neighbor. So, x is adjacent at least one of u or v in X^c. Also, note that u, v are adjacent in X c . Therefore, for any x, y ∈V(X^c) = V(X), there is path of length at most 3 in X^c through u, v, namely either x–u–v–y or x–v–u–y. Thus, the required result follows.

Connectedness in Directed graphs:

- Weakly connected: A digraph is weakly connected if it is connected as an undirected graph (a graph in which the direction of the edges is neglected).
- Unilaterally connected: If for any pair of nodes of the graph at least one of the nodes of the pair is reachable from the other.
- Strongly connected: If for any pair of nodes of the graph both the nodes of the pair are reachable from one to another.

From the definition every strongly connected graph is unilaterally connected and every unilaterally connected graph is weakly connected graph.

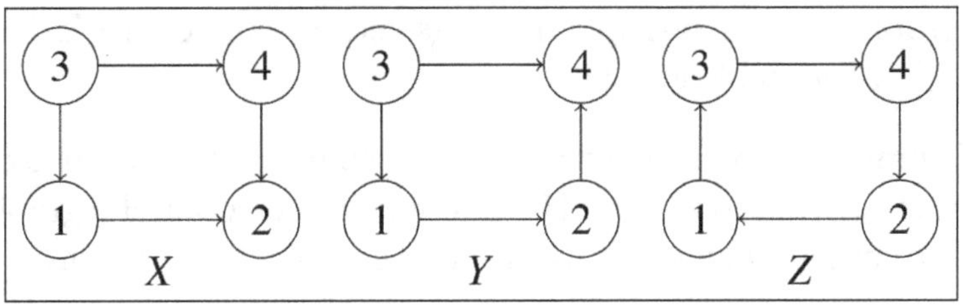

It is easy check that the graph Z is strongly connected; Y is unilaterally connected but not strongly connected, whereas X is weakly connected but not unilaterally connected.

Disconnecting Sets, Cut Sets

A disconnecting set in a connected graph X is a set of edges whose removal disconnects X.

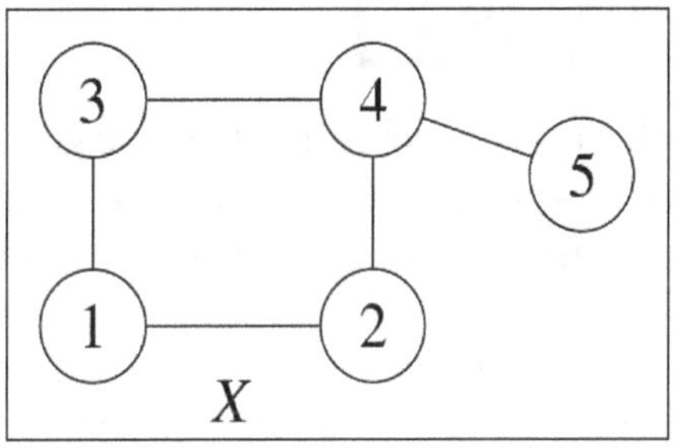

For example, some of the disconnecting sets are:

$$\{\{4,5\}\},\{\{1,2\},\{3,4\}\},\{\{1,3\},\{2,4\}\},\{\{1,2\},\{3,4\},\{4,5\}\}$$

Cut set is a disconnecting set having the property that no proper subset of which is a disconnecting set. Cut set with only one edge is called a bridge.

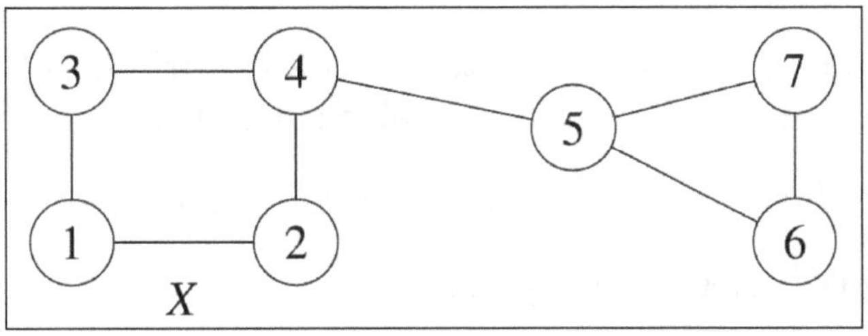

The edge {4,5} is a bridge. A graph may contain cut sets of different sizes.

- Edge connectivity: Let X be a connected graph. Then, its edge connectivity, denoted λ(X), equals.

  λ(X) = The number of elements in the smallest cutest.

  = Minimum number of edges required to disconnect the graph.

- Separating set: It is a set of vertices whose removal disconnects the graph.

Vertex connectivity of a connected graph X, denoted κ(X), equals the minimum number of vertices required to remove so that the graph becomes disconnect.

- A connected graph is said to be separable if its vertex connectivity is one. All other connected graphs are non-separable graphs.
- In a separable graph a vertex whose removal disconnects the graph is called a cut vertex or a cut node or an articulation point. [In a tree (connected acyclic graph) every vertex with degree greater than 1 is a cut vertex. So, every tree with three or more vertices is a separable graph].
- A separable graph consists of two or more non separable sub graphs.
- Each of the largest non-separable sub graph is called a block.

Representation of Graphs

In graph theory, a graph representation is a technique to store graph into the memory of computer. To represent a graph, we just need the set of vertices, and for each vertex the neighbors of the vertex (vertices which is directly connected to it by an edge). If it is a weighted graph, then the weight will be associated with each edge.

There are different ways to optimally represent a graph, depending on the density of its edges, type of operations to be performed and ease of use.

Adjacency Matrix

- Adjacency matrix is a sequential representation.
- It is used to represent which nodes are adjacent to each other. i.e. is there any edge connecting nodes to a graph.
- In this representation, we have to construct a nXn matrix A. If there is any edge from a vertex i to vertex j, then the corresponding element of A, a ^{i,j} = 1, otherwise a ^{i,j}= 0.
- If there is any weighted graph then instead of 1s and 0s, we can store the weight of the edge.

Consider the following undirected graph representation:

Undirected Graph Representation

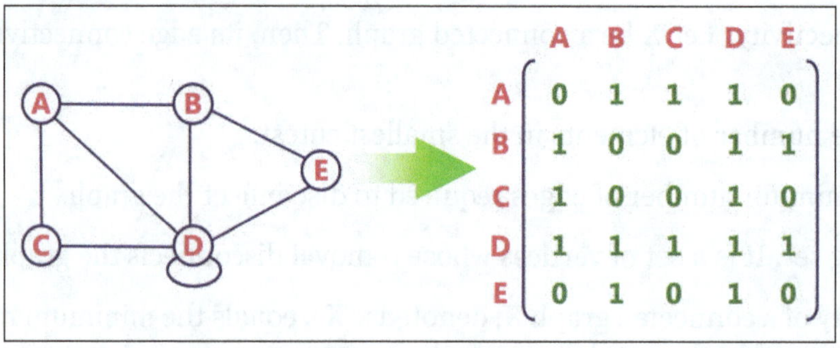

Directed Graph Representation

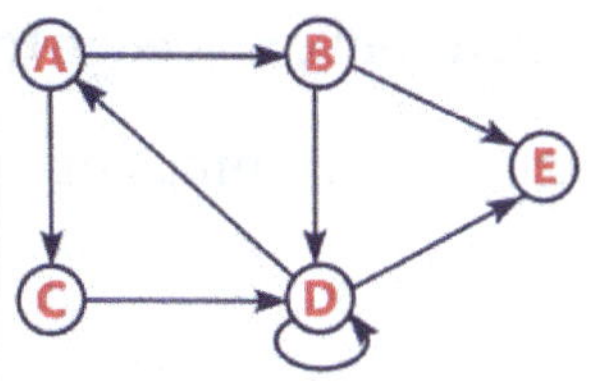

	A	B	C	D	E
A	0	1	1	0	0
B	0	0	0	1	1
C	0	0	0	1	0
D	1	0	0	1	1
E	0	0	0	0	0

In the above examples, 1 represents an edge from row vertex to column vertex, and 0 represents no edge from row vertex to column vertex.

Undirected Weighted Graph Representation

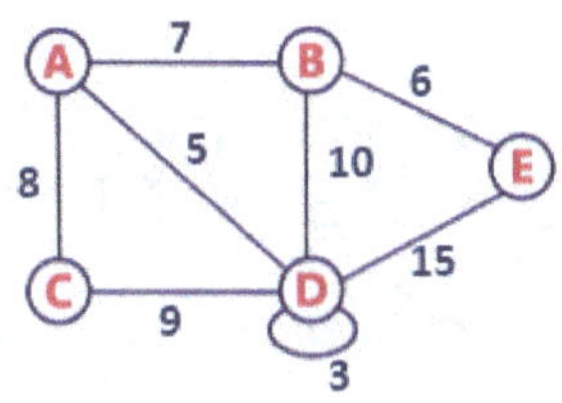

	A	B	C	D	E
A	0	7	8	5	0
B	7	0	0	10	6
C	8	0	0	9	0
D	5	10	9	3	15
E	0	6	0	15	0

- Pros: Representation is easier to implement and follow.
- Cons: It takes a lot of space and time to visit all the neighbors of a vertex, we have to traverse all the vertices in the graph, which takes quite some time.

Incidence Matrix

In Incidence matrix representation, graph can be represented using a matrix of size: Total number of vertices by total number of edges.

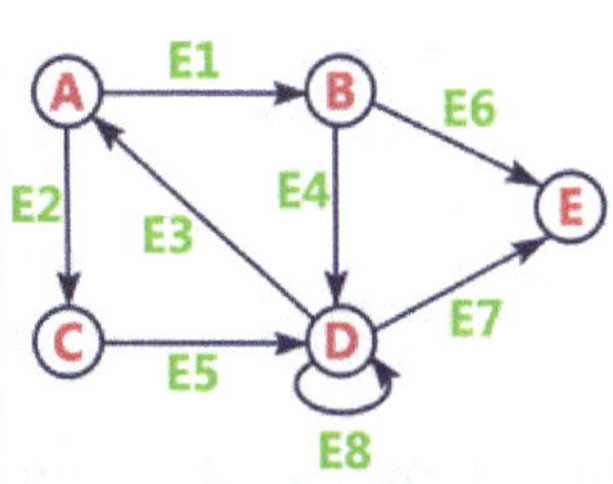

	E1	E2	E3	E4	E5	E6	E7	E8
A	1	1	-1	0	0	0	0	0
B	-1	0	0	1	0	1	0	0
C	0	-1	0	0	1	0	0	0
D	0	0	1	-1	-1	0	1	1
E	0	0	0	0	0	-1	-1	0

It means if a graph has 4 vertices and 6 edges, then it can be represented using a matrix of 4X6 class. In this matrix, columns represent edges and rows represent vertices. This matrix is filled with either 0 or 1 or -1.

Where,

- 0 is used to represent row edge which is not connected to column vertex.
- 1 is used to represent row edge which is connected as outgoing edge to column vertex.
- -1 is used to represent row edge which is connected as incoming edge to column vertex.

Adjacency List

- Adjacency list is a linked representation.
- In this representation, for each vertex in the graph, we maintain the list of its neighbors. It means, every vertex of the graph contains list of its adjacent vertices.
- We have an array of vertices which is indexed by the vertex number and for each vertex v, the corresponding array element points to a singly linked list of neighbors of v.

Let's see the following directed graph representation implemented using linked list:

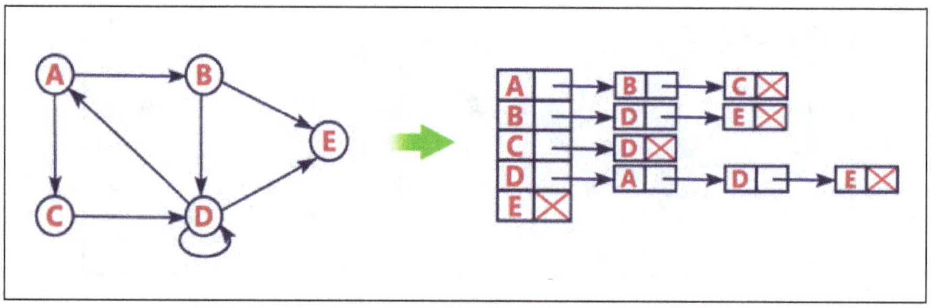

We can also implement this representation using array as follows:

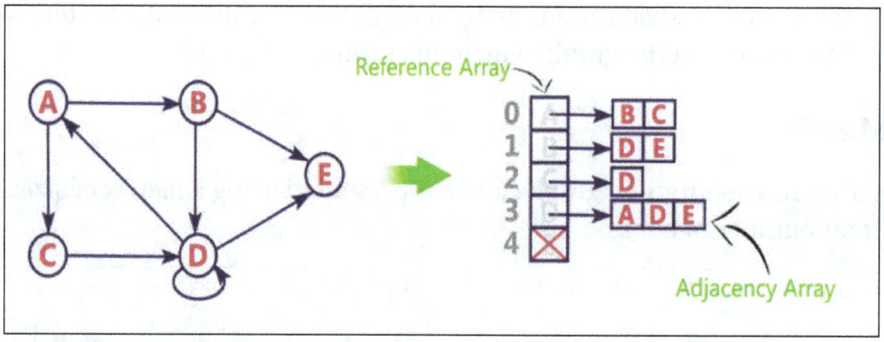

Pros:

- Adjacency list saves lot of space.
- We can easily insert or delete as we use linked list.
- Such kind of representation is easy to follow and clearly shows the adjacent nodes of node.

Cons:

- The adjacency list allows testing whether two vertices are adjacent to each other but it is slower to support this operation.

Isomorphism

Most properties of a graph do not depend on the particular names of the vertices. For example, although graphs A and B are technically different (as their vertex sets are distinct), in some very important sense they are the "same" graph. For example, both graphs are connected, have four vertices and three edges. However, graph C also has four vertices and three edges, and yet as a graph it seems different from the first two. Isomorphism is the idea that captures the kind of sameness that we recognize between A and B, and which distinguishes both of them from C.

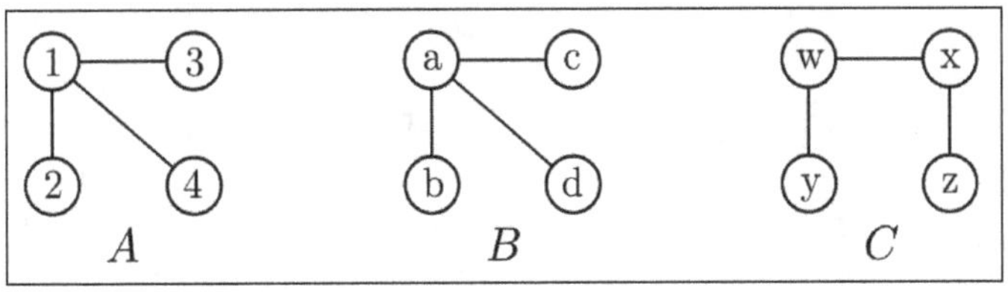

Two isomorphic graphs A and B and a non-isomorphic graph C; each have four vertices and three edges.

Definition: Two graphs G_1 and G_2 are isomorphic if there exists a matching between their vertices so that two vertices are connected by an edge in G_1 if and only if corresponding vertices are connected by an edge in G_2.

In figure, we can match the vertices of graph A with those of graph B in such a way: 1 is matched with a, 2 with b, 3 with c, and 4 with d. An edge connects 1 and 3 in the first graph, and so an edge connects a and c in the second graph. Likewise, no edge connects 3 and 4 in the first graph, and so no edge connects c and d in the second graph. Regarding the two graphs in figure 10, we can write A ≅ B to denote this isomorphism. Although we matched vertices of A with those of B in one particular way, there could be several ways to do. For example, we could match 1 with a, 2 with c, 3 with d, and 4 with b; there are several other ways to do this. We often use the symbol ≅ to denote isomorphism between two graphs, and so would write A ≅ B to indicate that A and B are isomorphic.

Although graphs A and B are isomorphic, i.e., we can match their vertices in a particular way, graph C is not isomorphic to either of A or B. As hard as we try, we will fail to find a matching between vertices of A, for example, and those of C that maintain edge-connections between corresponding vertices. We can write A C to indicate that A and C are not isomorphic.

Graph theorists are primarily interested in properties of graphs that do not change when vertices are relabeled; sometimes they will discuss "properties that are invariant under isomorphism's" which conveys this idea. if $G_1 \cong G_2$, then many properties of G_1 and G_2 must be the same. For example, the number of vertices and edges in the two graphs must be identical. G_1 is connected if and only G_2 is connected. G_1 is k-regular if and only if G_2 is k-regular. G_1 is bipartite if and only if G_2 is bipartite. The numbers of vertices with degree 0, 1, 2, etc. must be identical. None of the properties change when vertices are relabeled. Conversely, if two graphs G_1 and G_2 differ with respect to any of these properties, then we can know that G_1 and G_2 are not isomorphic.

Many properties of individual vertices also do not change "under isomorphisms", or relabeling of the vertices. For example, if vertex u in graph G_1 can be matched with vertex v in G_2, then we must have deg(u) = deg(v). If deg(u) 6 ≠ deg(v), then we cannot match up the two vertices. Determining whether two graphs are isomorphic is not always an easy task. For graphs with only several vertices and edges, we can often look at the graph visually to help us make this determination.

A relabeling of vertices of a graph is isomorphic to the graph itself. Consider the three isomorphic graphs illustrated in figure. The first two graphs illustrate a change of using letters to using numbers to label the graphs. The second pair of graphs are also isomorphic as only the labels were changed. We can match vertices in the second graph with those in the third graph to satisfy the isomorphism requirements. Another way to think about graph isomorphism is by removing all vertex labels from two graphs. It is clear for these examples that all three graphs are then identical.

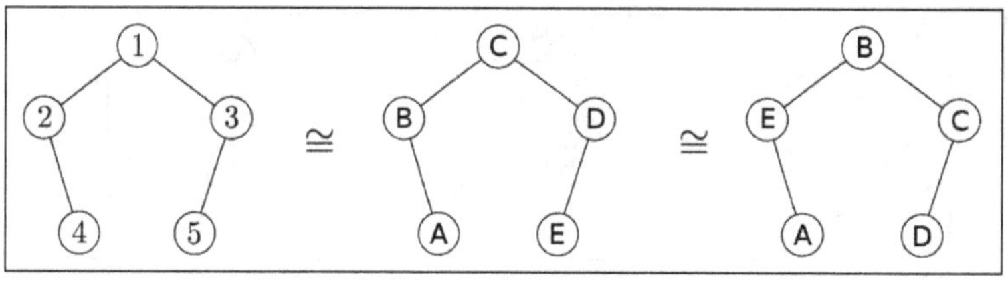

Three isomorphic graphs.

Like relabeling, moving around vertices also does not change important graph properties. The two graphs illustrated below are isomorphic since edges connected in one are also connected in the other. In fact, not only are the graphs isomorphic to one another, but they are in fact identical. Each vertex in one graph is matched to itself in the other graph.

The figure below illustrates another pair of isomorphic graphs. Although the graphs have slightly different shapes from one another, we can still find a 1-1 matching between the vertices so that if pairs of vertices are connected by an edge in one graph, then corresponding vertices will be connected in the other. The same matching given above (a1, b2, c3, d4) will still work here, even though we have moved the vertices around. It is worth noting that several other matching's would also work. For example a1, b3, c4, d2 would also be a good matching. In fact, as long as a is matched with 1, then b, c, and d can be matched in any order with 2, 3, and 4.

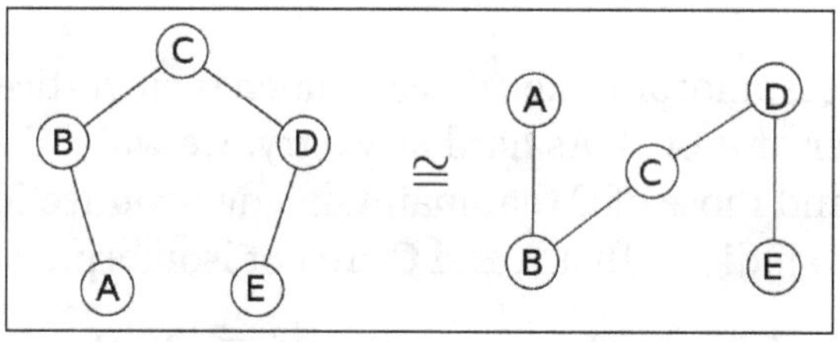

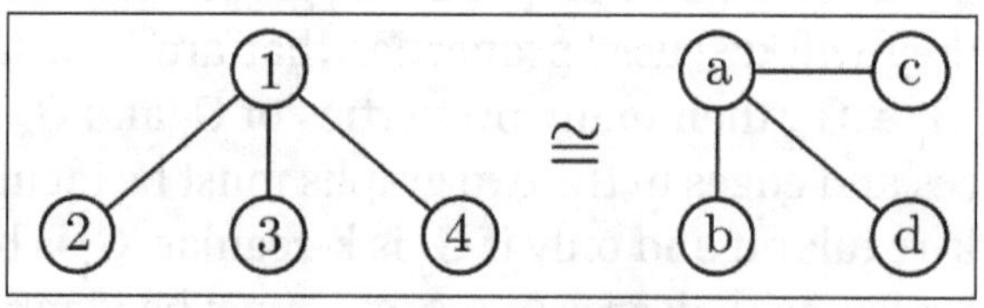

Two isomorphic graphs.

Any two cliques K_n and K_m are isomorphic if any only if $n = m$. If $n \neq m$, then it is clear that we cannot have a 1-1 matching between the vertices of the two graphs, because there will be more vertices in one graph than in the other. If $n = m$ then any matching will work, since all pairs of distinct vertices are connected by an edge in both graphs. Any matching of the vertices will ensure the isomorphism definition is satisfied.

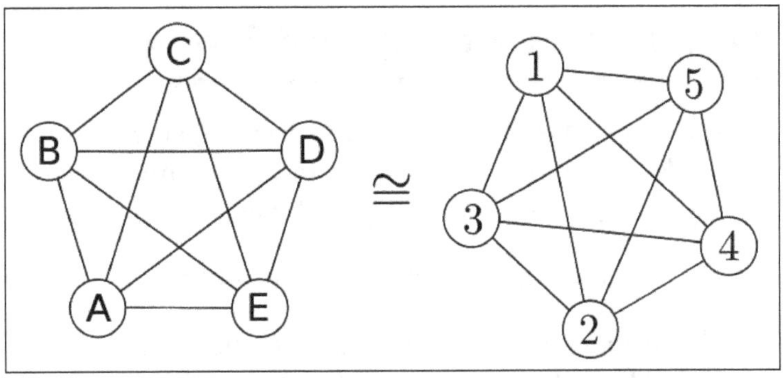

Two complete graphs on five vertices; they are isomorphic.

Just because two graphs have the same number of vertices and edges does not mean that they are isomorphic. In fact, even if the degrees of all vertices are identical, still the two graphs can be non-isomorphic. Although $G_1 \cong G_2$ (we can imagine deforming either into the other), G_2 and G_3 are not isomorphic. No matter how we relabel the vertices of G_2, it will remain a connected graph. Likewise, no matter how we relabel the vertices of a G_3, it will remain unconnected. Two graphs that are isomorphic must both be connected or both disconnected.

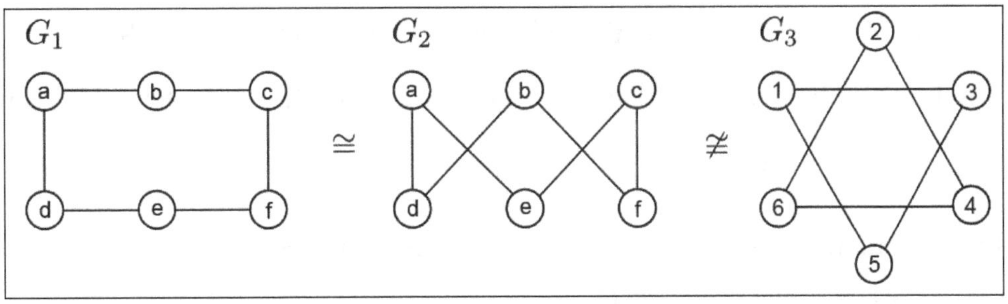

Three 2-regular graphs on six vertices; the first two are isomorphic; the third one is not.

Below are two complete graphs, or cliques, as every vertex in each graph is connected to every other vertex in that graph. These two graphs are isomorphic since they have the same number of vertices. The two drawing here, however, highlight a particularly interesting feature of certain graphs. We have already seen that we can generally draw graphs in many different ways without changing their overall structure. So long as we don't disconnect any vertices that are connected to each other, and so long as we don't attach any vertices that were previously disconnected, we are free to move the vertices and edges as we please. In the first way we drew the graph, two of its edges (AC and BD) crossed one another. However, in the second drawing, of an isomorphic graph, we were able to draw the graph in such a way that no two edges cross. In other words, there are some graphs that can be drawn without edges crossing.

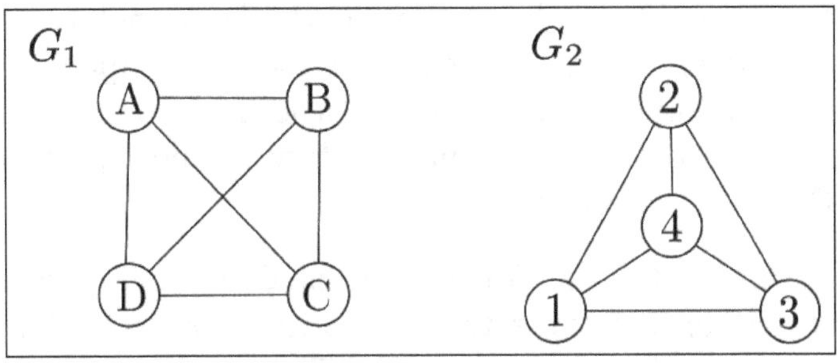

Two complete graphs on four vertices; they are isomorphic.

Graph Coloring and Homomorphism

A proper coloring of a graph G is an assignment of colors to V (G) such that no two adjacent vertices have the same color. A graph's chromatic number is the minimum number of colors needed to properly color the graph. We say that a graph is n-colorable if it can be properly colored with n colors.

Theorem: A graph G is r-colorable $\Leftrightarrow$ there exists a homomorphism from G to K_r.

Proof: ($\Rightarrow$) Assume G can be properly colored with r colors labeled {1, 2, . . . , r}. We label a generic vertex v of G by v_i if it is colored by i where $1 \leq i \leq r$. Define a mapping $h : G \rightarrow K_r$ by $h(v_i) = k_i$ where ki is the i th vertex in K_r (without loss of generality we can impose an ordering on the vertices of K_r). We claim that this map is a graph homomorphism: Let a, b $\in$ V (G) and assume a ~ b (a is adjacent to b),

- $\Rightarrow$ a and b do not have the same color so $a = a_i$ and $b = b_j$ where $i \neq j$.
- $\Rightarrow h(a) = h(a_i) = k_i 6= k_j = h(b_j) = h(b)$ so $h(a) \neq h(b)$.
- $\Rightarrow h(a) \sim h(b)$ (the image of h is a complete graph).

Therefore, $a \sim b \Rightarrow h(a) \sim h(b)$ showing that h is a homomorphism from G to K_r.

Let $h : G 7\rightarrow K_r$ be a graph homomorphism. For a given $y \in V(K_r)$ define the set $h^{-1}(y) \subseteq V(G)$ to be,

$$h^{-1}(y) = \{x \in V(G) \mid h(x) = y\}$$

It is important to remember that these sets might be empty. Let $a, b \in h^{-1}(y)$,

- $\Rightarrow h(a) = h(b) = y \left(a, b \in h^{-1}(y)\right)$.
- $\Rightarrow h(a) \nsim h(b)$ (a vertex cannot be adjacent to itself).
- $\Rightarrow a \nsim b$ (contrapositive of $a \sim b \Rightarrow h(a) \sim h(b)$).

So, for every $y \in V(K_r)$ the corresponding $h^{-1}(y)$ is an independent subset of V (G). Therefore, every element of $h^{-1}(y)$ can be colored using the same color. There are r vertices of K_r, and the sets $h^{-1}(y)$ form a partition of the vertices of G (partition comes from properties of a map, we can't assign one input to two different outputs and every input must map to some element of the codomain), so there are at most r colors needed to properly color G.

This theorem gives us a concrete mathematical interpretation of coloring a graph, but the above result answers much more than the simple question of whether or not a graph is n-colorable. As a simple consequence of this theorem, the chromatic number of a graph G is the smallest integer r such that there exists a homomorphism from G to K_r. In addition to this small logical step, a number of facts become apparent from the following corollary. A coloring of a graph G is precisely a homomorphism from G to some complete graph.

If we view a homomorphism $h : G \rightarrow K_r$ for some graph G as an assignment of colors to the vertices of G, then h directly tells us how to create this coloring. For any $a \in V(G)$, if $h(a) = k_i$ then we simply assign color i from a set of r colors to vertex a. This process gives us a unique coloring of

G for every h. If we want to know how many proper colorings of G we can create with r colors, we simply count the number of distinct homomorphisms from G to K_r.

Consider the following coloring of the Petersen graph P represented as a homomorphism from P $\rightarrow K_3$ where vertices of P map to their corresponding color in K_3. It should be fairly obvious that this map is indeed a graph homomorphism.

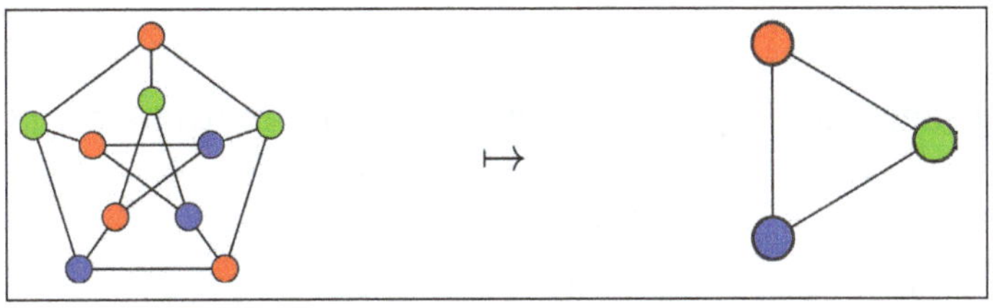

At this point, it is natural to wonder why graph coloring is important enough to warrant such a careful study. One would hope that there are more applications of this process than simply assigning colors to vertices, and indeed there are. Consider the problem of scheduling final exams such that no student is scheduled to take two exams during the same period. We can construct a graph G which has classes as its vertices, where two vertices are adjacent if there is a student taking both classes. We now have a graphical representation G of the current state our classes, and we wish to assign final periods to the vertices of G such that no two adjacent vertices have the same period assignment (otherwise this would constitute a scheduling conflict). Finding such a scheduling of our classes is equivalent to finding a proper coloring of G, where each color represents a different exam period.

We can assign frequencies to radio stations without causing interference arising from two geographically close radio stations having the same frequency. We can help an airline company assign pilots to their flights where each pilot is a color, each flight is a vertex, and each edge represents two flights that cannot be piloted by the same person. We can make as many examples of these types of problems as we want by simply inventing probable scenarios of this form. You might be tempted to say that these scenarios are all the same problem worded differently, but instead think of them as a family of problems that can all be solved using the same technique.

Our family of problems solvable by graph coloring is indeed expansive, but it feels quite constrained because all of our solutions only consist of assigning n colors to a given graph. If we want to ensure that no student is scheduled to take 2 exams in consecutive periods, we must change the desired homomorphic image from a complete graph to a more restrictive graphical representation of our optimal schedule. If we have 6 final periods with exams already scheduled such that no student must take consecutive exams, we can model this schedule as:

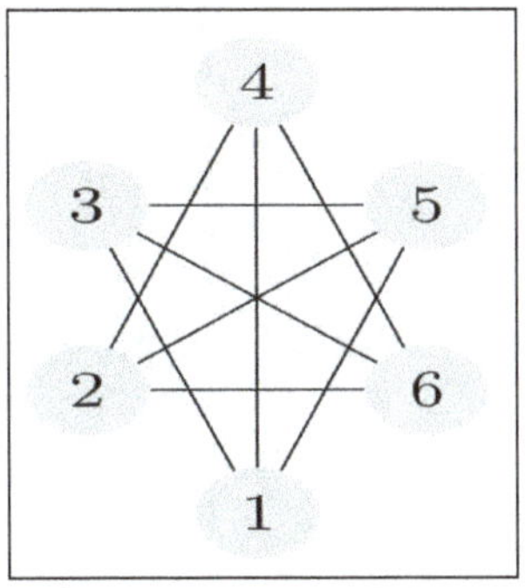

Where each vertex is a final exam period and there exists an edge between vertices if a student can be scheduled to take an exam in both time slots. Once we have found a graphical representation of a schedule satisfying the constraints of our problem, we must ask if such a scheduling is possible, and if so how can we achieve it. To answer these questions we can again construct a graphical representation G of our current classes (classes are vertices of G while edges represent classes with mutual students) and find a homomorphism from G to our desired schedule represented by a graph H.

Consider the following schedule assignment as a graph homomorphism h : G → H where,

$G =$

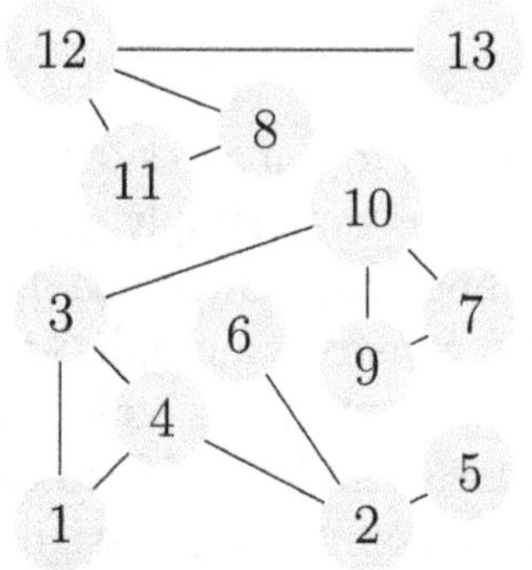

$h:$

$1 \mapsto 1$, $2 \mapsto 2$, $3 \mapsto 3$, $4 \mapsto 5$, $5 \mapsto 6$, $6 \mapsto 5$, $7 \mapsto 2$, $8 \mapsto 1$, $9 \mapsto 3$, $10 \mapsto 6$, $11 \mapsto 5$, $12 \mapsto 3$, $13 \mapsto 6$

$H =$

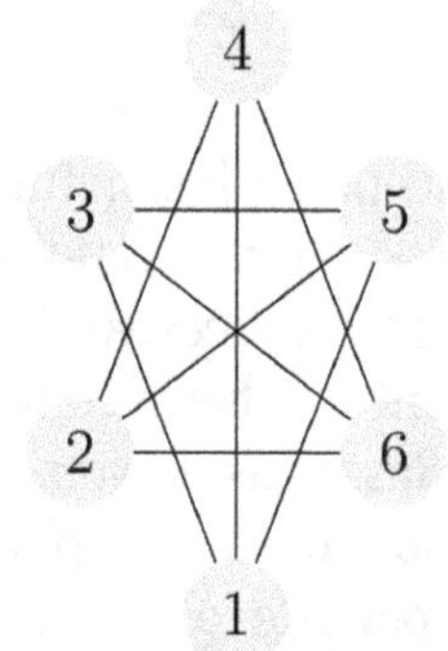

This homomorphism creates a scheduling that satisfies our constraints because two adjacent vertices in G (i.e. two classes that share at least one student) cannot be mapped to vertices of the form (i, i + 1 mod 6) due to the definition of graph homomorphisms. This is equivalent to saying that no two classes that share at least one student can have their final exams scheduled in consecutive time slots. Our original requirement that no two classes that share students can be scheduled at the same time is still satisfied, because adjacent vertices in G must be mapped to adjacent in H and no vertex is adjacent to itself.

Period 4 is never used, showing that that H does not represent an optimal schedule of final exam periods. With these specific constraints on our desired schedule, we must find the smallest n for which there exists a homomorphism from G to H_n. We define H_n to be the subgraph of K_n with edges (i, i+1 mod n) removed. It is useful to think of this definition of H_n as if we have labeled the vertices of K_n as {1, 2, . . . , n} in increasing order around the perimeter of K_n. However this interpretation is not necessary as all labelings of K_n are isomorphic.

Lunch break (between periods 3 and 4)

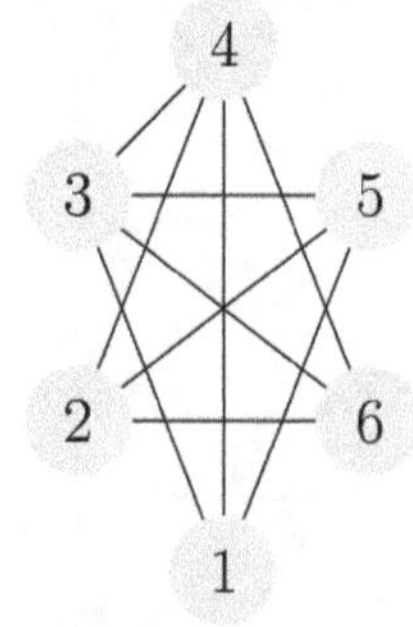

Dinner break (between periods 5 and 6)

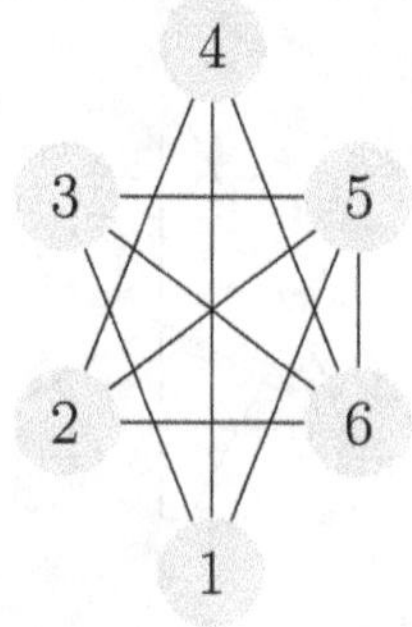

Both

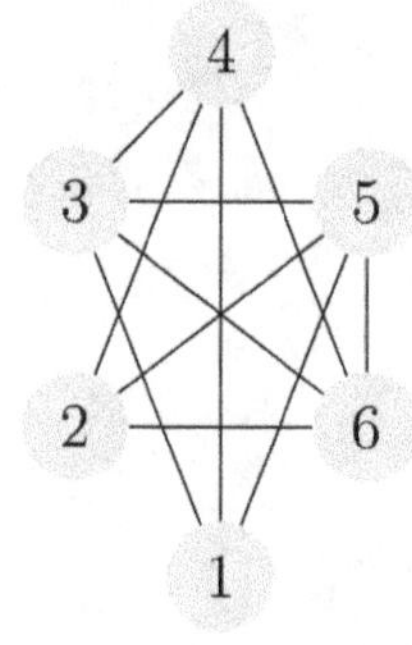

It now becomes easy to modify our constraints on our optimal schedule. We can add and remove edges and vertices to our graph H (the graphical representation of our desired schedule) depending on our new constraints.

If our desired schedule has 6 periods, again with the constraint that no student is scheduled to take two exams consecutively, we can modify this to include a lunch break (so exams can be schedule on either side of this break without conflict), or a dinner break, or both.

To drive home the point, we arrive at our final example.

Homomorphisms

To ease notation, we write $G \rightarrow H$ if there exists a graph homomorphism from G to H. A clique of a graph G is a subgraph of G which is complete. We say that a graph has clique number n, denoted $\omega(G)$, if the largest clique of G has n vertices. Just as $\chi(G)$ is the smallest r such that $G \rightarrow K_r$, $\omega(G)$ is the smallest n such that $K_n \rightarrow G$. We state this as a theorem for consistency.

Theorem: For any graph G, $\omega(G) = r \Leftrightarrow r$ is the largest integer such that $K_r \rightarrow G$.

Proof: If G has a clique of size n labeled C, then any isomorphism from K_n to C induces a homomorphism from K_n to G (the fact that C is isomorphic to K_n should be obvious, considering they are both complete graphs on n vertices). Conversely, if $K_n \rightarrow G$ then each vertex of K_n must have a unique image in G, where all of these images must be connected in order to preserve the edges of K_n. These images must then form a clique of size n in G. Given this necessary and sufficient condition, it directly follows that $\omega(G) = r \Leftrightarrow K_r \rightarrow G$ where r is maximal.

The chromatic number of a graph G, denoted $\chi(G)$, as the minimum number of colors needed to properly color G. The following theorem describes a useful relationship between both $\chi(G)$ and $\chi(H)$, and $\omega(G)$ and $\omega(H)$ respectively, where H is the codomain of a homomorphism from G.

Theorem: If $G \rightarrow H$ then $\omega(G) \leq \omega(H)$ and $\chi(G) \leq \chi(H)$.

Proof: The composition of homomorphisms is again a homomorphism, i.e. $G \rightarrow H \rightarrow K$ implies $G \rightarrow K$. Given this fact,

$K\omega(G) \rightarrow G \rightarrow H$ implies $K\omega(G) \rightarrow H$, so $\omega(H)$ is at least $\omega(G)$, and $G \rightarrow H \rightarrow K\chi_{(H)}$ implies $G \rightarrow K\chi_{(H)}$ so $\chi(G)$ is at most $\chi(H)$.

Even with these simple results, we are beginning to make significant process in determining if, given graphs G and H, there is no homomorphism from G to H. We can compute $\omega(G)$ and $\chi(G)$ for any graph G, although these computations might be difficult. Comparing our results for G and H we can immediately tell if G does not map homomorphically to H. If $G \leftarrow H$ and $H \leftarrow G$ we say that $G \equiv H$. '$\equiv$' produces an equivalence relation. The relation is symmetric because the identity map from G to G is an isomorphism and thus a homomorphism, transitive because composition of homomorphisms is again a homomorphism, and reflexive by nature of our definition. This equivalence relation forms equivalence classes of graphs, which we can use to solve our decision problem.

Since we are now dealing with equivalence classes, it is natural to ask if there are canonical representatives of these classes. We define a core of a graph G to be the element of [G] (homomorphism equivalence class of G) with the least number of vertices. Given this definition, we get some interesting results involving these cores.

- G is a core ⇔ every endomorphism of G is an automorphism.
- Any graph (or homomorphism equivalence class) has a unique core unto isomorphism.

These results are certainly nontrivial, G is a core ⇔ every endomorphism of G is an automorphism gives sufficient proofs of these facts and a discussion of the topic of cores and homomorphism equivalence classes in the context of classifying graphs, Instead, let us examine why the above results might prove useful.

Since the number of vertices of the core of a graph G must be less than or equal to |V (G)|, we assume it is more computationaly simple to find homomorphism from this core to a desired H than using our original G. If we denote the core of G by C(H), it should be clear that G → H if and only if C(G) → H. Therefore, we can refine our search for G → H for all graphs homomorphically equivalent to G. This is useful because cores have specific properties that they do not share with generic graphs.

Trees

A graph with no cycles is called acyclic. An acyclic graph is called a forest. A forest having only one component is called a tree, i.e., a tree is a connected acyclic graph. A tree T with only one vertex is called a trivial tree and is called non trivial, otherwise. The following are 6 non isomorphic trees on 6 vertices.

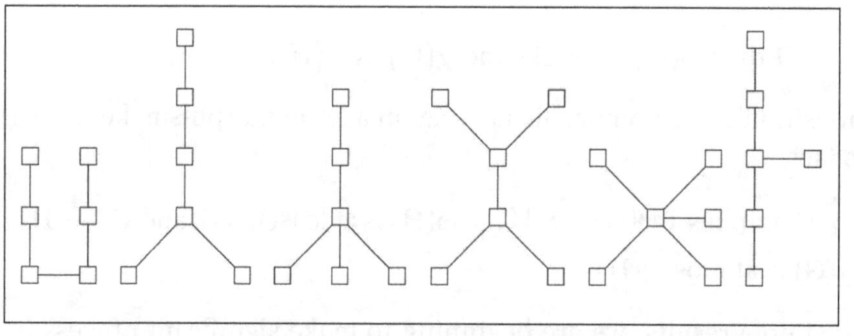

Number of vertices	Number of non-isomorphic trees
1	1
2	1
3	1
4	2
5	3
6	6

7	11
8	23
9	47
10	106
11	235
12	551
13	1301
14	3159
15	7741
16	19320

Let X be a simple graph on n vertices. Then, the following statements are equivalent.

1. X is a tree (connected acyclic graph).
2. There is a unique path between any two vertices of X.
3. X is connected and has exactly n−1 edges.
4. X contains no cycles and has n−1 edges.
5. X is minimally connected, i.e., every edge is a bridge.

Proof: 1 ⇒ 2 Let X be a tree. Then, X is connected and there is a path between any two vertices. Suppose there exists at least two distinct paths between a pair of vertices, say $u, v \in V(X)$. Then, the union of these two paths produces a cycle. A contradiction.

2 ⇒ 1 Let X be a graph with a unique path between any two vertices. Then, clearly X is connected. On the contrary assume that there is a cycle in the graph. Then, there exists at least two vertices on this cycle that have two distinct paths between them. Hence, a contradiction.

2 ⇒ 3 Let X be a connected graph with a unique path between any two vertices, then clearly X is connected. We will show that X has n−1 edges by induction on n, the number of vertices. The result is clearly true if n = 1,2,3. Now, assume that the result is true for all graphs containing less than k vertices and let X be a graph with k vertices. Since there is a unique path between any two vertices, if we remove any edge e from the graph X, then $X \setminus \{e\}$ has k vertices, but has two connected components, say X_1 and X_2 with vertices $k_1 \geq 1$ and $k_2 \geq$, respectively, with $k_1 + k_2 = k$. Thus, by induction X_1 and X_2 contain $k_1 - 1$ and $k_2 - 1$ edges, respectively. Thus, the number of edges in X is $(k_1 - 1) + (k_2 - 1) + 1 = k - 1$.

3 ⇒ 4 Suppose X is connected and has exactly n−1 edges. We have to show that X has no cycles. On the contrary assume that X has at least one cycle. Then, if we delete any edge from the cycle, the new graph is still connected and hence it has at least n−1 edges. Thus, we arrive at a contradiction.

4 ⇒ 5 Suppose X has k components. We show that k = 1. Since X is acyclic, each component is acyclic. That is, each component is a tree and hence (use 1 ⇔ 2 ⇒ 3) X contains n−k edges. Thus, k = 1 and therefore, X is a tree. Hence, there is a unique path between any two vertices in X (use 1 ⇔ 2). Hence, every edge is bridge.

5 ⇒ 1 Let X be minimally connected. Then, X is connected and there exists a unique path between any two vertices. Thus, X it tree (use 1 ⇔ 2).

References

- Graph-theory-fundamentals: tutorialspoint.com, Retrieved 03, February 2020
- Graph-theory-types-of-graphs: javatpoint.com, Retrieved 21, July 2020
- Graph-theory-graph-representations: javatpoint.com, Retrieved 05, March 2020
- Davis-homomorphism-ups-434: ups.edu, Retrieved 14, August 2020

3

Sets, Relations and Functions

A set refers to an unordered collection of objects. These objects are called members of the set. A few types of sets are finite and countably infinite sets. A function is used to assign one member of a related set to each member of a particular set. The topics elaborated in this chapter will help in gaining a better perspective about sets, relations and functions.

A set is an unordered collection of different elements. A set can be written explicitly by listing its elements using set bracket. If the order of the elements is changed or any element of a set is repeated, it does not make any changes in the set.

Some Example of Sets

- A set of all positive integers.
- A set of all the planets in the solar system.
- A set of all the states in India.
- A set of all the lowercase letters of the alphabet.

Representation of a Set

Sets can be represented in two ways – Roster or Tabular Form and Set Builder Notation.

Roster or Tabular Form

The set is represented by listing all the elements comprising it. The elements are enclosed within braces and separated by commas. Example:

- Set of vowels in English alphabet, $A = \{a, e, i, o, u\}$.
- Set of odd numbers less than 10, $B = \{1, 3, 5, 7, 9\}$.

Set Builder Notation

The set is defined by specifying a property that elements of the set have in common. The set is described as $A = \{x : p(x)\}$. Example:

- The set $\{a, e, i, o, u\}$ is written as,

  $A = \{x : x \text{ is a vowel in English alphabet}\}$
- The set $\{1, 3, 5, 7, 9\}$ is written as,

  $B = \{x : 1 \le x < 10 \text{ and } (x \% 2) \ne 0\}$

If an element x is a member of any set S, it is denoted by $x \in S$ and if an element y is not a member of set S, it is denoted by $y \notin S$.

Example: If $S = \{1, 1.2, 1.7, 2\}, 1 \in S$ but $1.5 \notin S$.

Some Important Sets

- N: The set of all natural numbers = {1, 2, 3, 4,.....}.
- Z: The set of all integers = {....., −3, −2, −1, 0, 1, 2, 3,.....}.
- Z^+: The set of all positive integers.
- Q: The set of all rational numbers.
- R: The set of all real numbers.
- W: The set of all whole numbers.

Cardinality of a Set

Cardinality of a set S, denoted by |S|, is the number of elements of the set. The number is also referred as the cardinal number. If a set has an infinite number of elements, its cardinality is ∞.

Example: $|\{1,4,3,5\}| = 4, |\{1,2,3,4,5,\ldots\}| = \infty$

If there are two sets X and Y:

- $|X| = |Y|$ denotes two sets X and Y having same cardinality. It occurs when the number of elements in X is exactly equal to the number of elements in Y. In this case, there exists a bijective function 'f' from X to Y.
- $|X| \leq |Y|$ denotes that set X's cardinality is less than or equal to set Y's cardinality. It occurs when number of elements in X is less than or equal to that of Y. Here, there exists an injective function 'f' from X to Y.
- $|X| < |Y|$ denotes that set X's cardinality is less than set Y's cardinality. It occurs when number of elements in X is less than that of Y. Here, the function 'f' from X to Y is injective function but not bijective.
- If $|X| \leq |Y|$ and $|X| \geq |Y|$ then $|X| = |Y|$. The sets X and Y are commonly referred as equivalent sets.

Types of Sets

Sets can be classified into many types. Some of which are finite, infinite, subset, universal, proper, singleton set, etc.

1. Finite Set: A set which contains a definite number of elements is called a finite set.

Example: $S = \{x \mid x \in N \text{ and } 70 > x > 50\}$

2. Infinite Set: A set which contains infinite number of elements is called an infinite set.

Example: $S = \{x \mid x \in N \text{ and } x > 10\}$

3. Subset: A set X is a subset of set Y (Written as $X \subseteq Y$) if every element of X is an element of set Y.

Example: Let, $X = \{1,2,3,4,5,6\}$ and $Y = \{1,2\}$. Here set Y is a subset of set X as all the elements of set Y is in set X. Hence, we can write $Y \subseteq X$.

Example: Let, $X = \{1,2,3\}$ and $Y = \{1,2,3\}$. Here set Y is a subset (Not a proper subset) of set X as all the elements of set Y is in set X. Hence, we can write $Y \subseteq X$.

4. Proper Subset: The term "proper subset" can be defined as "subset of but not equal to". A Set X is a proper subset of set Y (Written as $X \subset Y$) if every element of X is an element of set Y and $|X| < |Y|$.

Example: Let, $X = \{1,2,3,4,5,6\}$ and $Y = \{1,2\}$. Here set $Y \subset X$ since all elements in Y are contained in X too and X has at least one element is more than set Y.

3. Universal Set: It is a collection of all elements in a particular context or application. All the sets in that context or application are essentially subsets of this universal set. Universal sets are represented as U.

Example: We may define U as the set of all animals on earth. In this case, set of all mammals is a subset of U, set of all fishes is a subset of U, set of all insects is a subset of U, and so on.

4. Empty Set or Null Set: An empty set contains no elements. It is denoted by $\varnothing$. As the number of elements in an empty set is finite, empty set is a finite set. The cardinality of empty set or null set is zero.

Example: $S = \{x | x \in N \text{ and } 7 < x < 8\} = \varnothing$

5. Singleton Set or Unit Set: Singleton set or unit set contains only one element. A singleton set is denoted by {s}.

Example: $S = \{x | x \in N,\ 7 < x < 8 < 9\} = \{8\}$

6. Equal Set: If two sets contain the same elements they are said to be equal.

Example: If A = {1, 2, 6} and B = {6, 1, 2}, they are equal as every element of set A is an element of set B and every element of set B is an element of set A.

7. Equivalent Set: If the cardinalities of two sets are same, they are called equivalent sets.

Example: If A = {1, 2, 6} and B = {16, 17, 22}, they are equivalent as cardinality of A is equal to the cardinality of B. i.e. $|A| = |B| = 3$.

8. Overlapping Set: Two sets that have at least one common element are called overlapping sets. In case of overlapping sets:

- $n(A \cup B) = n(A) + n(B) - n(A \cap B)$

- $n(A \cup B) = n(A - B) + n(B - A) + n(A \cap B)$
- $(A) = n(A - B) + n(A \cap B)$
- $n(B) = n(B - A) + n(A \cap B)$

Example: Let, A = {1, 2, 6} and B = {6, 12, 42}. There is a common element '6', hence these sets are overlapping sets.

9. Disjoint Set: Two sets A and B are called disjoint sets if they do not have even one element in common. Therefore, disjoint sets have the following properties:

- $n(A \cup B) = \varnothing$
- $n(A \cup B) = n(A) + n(B)$

Example: Let, A = {1, 2, 6} and B = {7, 9, 14}, there is not a single common element hence these sets are overlapping sets.

Venn Diagrams

Venn diagram, invented in 1880 by John Venn, is a schematic diagram that shows all possible logical relations between different mathematical sets.

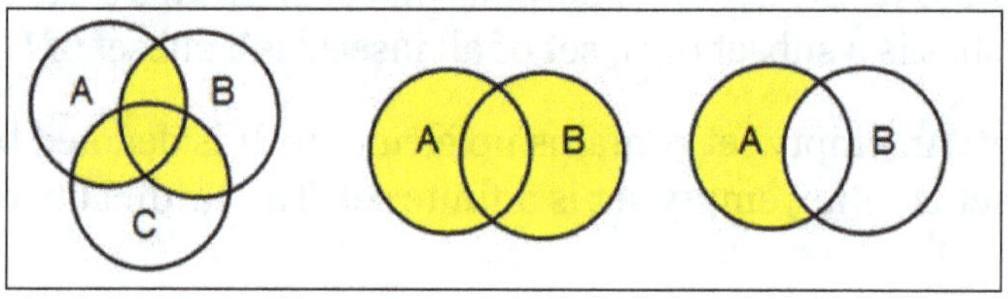

Set Operations

The union of two sets is a set containing all elements that are in A or in B (possibly both). For example, {1, 2} ∪ {2, 3} = {1, 2, 3}. Thus, we can write $x \in (A \cup B)$ if and only if $(x \in A)$ or $(x \in B)$. $A \cup B = B \cup A$. In figure below, the union of sets A and B is shown by the shaded area in the Venn diagram.

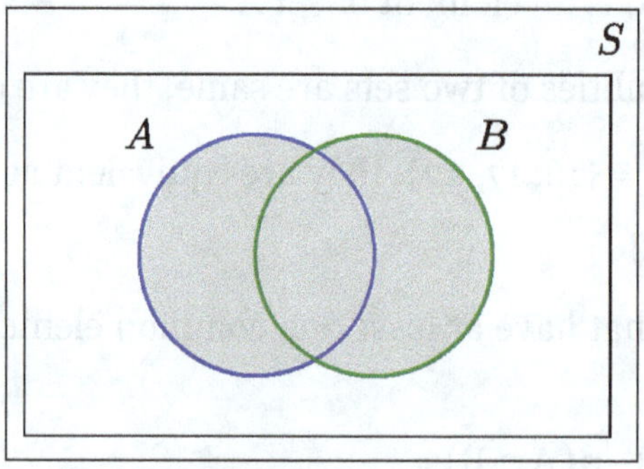

The shaded area shows the set $B \cup A$.

Similarly we can define the union of three or more sets. In particular, if $A_1, A_2, A_3, \ldots, A_n$ are n sets, their union $A_1 \cup A_2 \cup A_3 \ldots \cup A_n$ is a set containing all elements that are in at least one of the sets. We can write this union more compactly by,

$$\bigcup_{i=1}^{n} A_i.$$

For example, if $A_1 = \{a,b,c\}, A_2 = \{c,h\}, A_3 = \{a,d\}$, then $\bigcup_i A_i = A_1 \cup A_2 \cup A_3 = \{a,b,c,h,d\}$. We can similarly define the union of infinitely many sets $A_1 \cup A_2 \cup A_3 \cup \ldots$.

The intersection of two sets A and B, denoted by $A \cap B$, consists of all elements that are both in A and B. For example, $\{1,2\} \cap \{2,3\} = \{2\}$. In figure below, the intersection of sets A and B is shown by the shaded area using a Venn diagram.

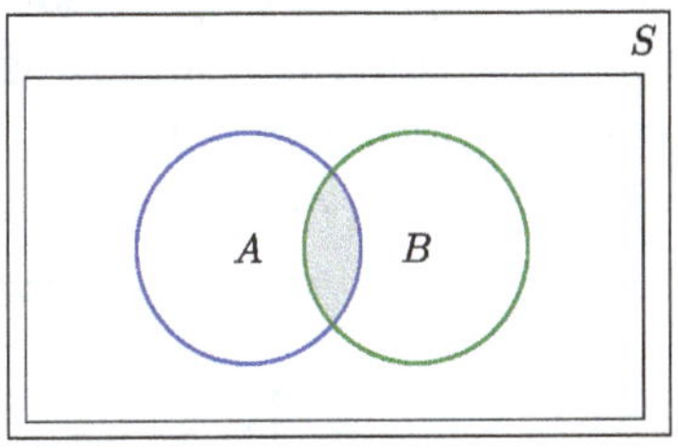

The shaded area shows the set $B \cap A$.

More generally, for sets $A_1, A_2, A_3, \ldots$, their intersection $\bigcap_i A_i$ is defined as the set consisting of the elements that are in all A_i's. Figure below shows the intersection of three sets.

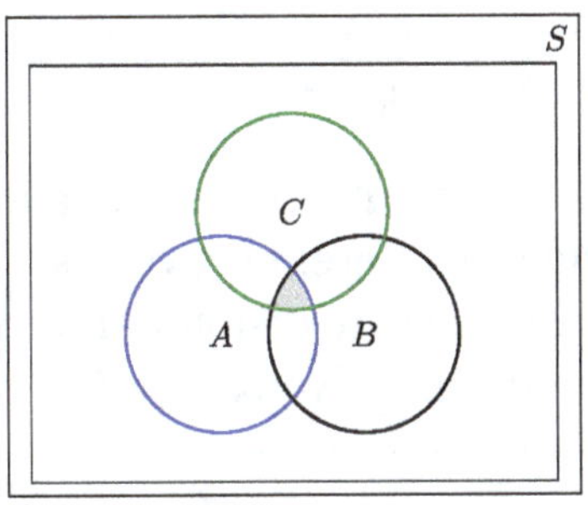

The shaded area shows the set $A \cap B \cap C$.

The complement of a set A, denoted by A^c or $\bar{A}$, is the set of all elements that are in the universal set S but are not in A. In figure below, $\bar{A}$ is shown by the shaded area using a Venn diagram.

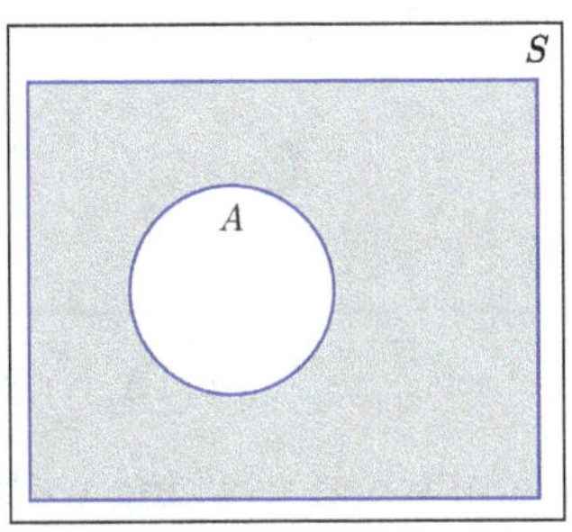

The shaded area shows the set $\bar{A} = A^c$.

The difference (subtraction) is defined as follows. The set A – B consists of elements that are in A but not in B. For example if A = {1, 2, 3} and B = {3, 5}, then A – B = {1, 2}. In figure below, A – B is shown by the shaded area using a Venn diagram. $A - B = A \cap B^c$.

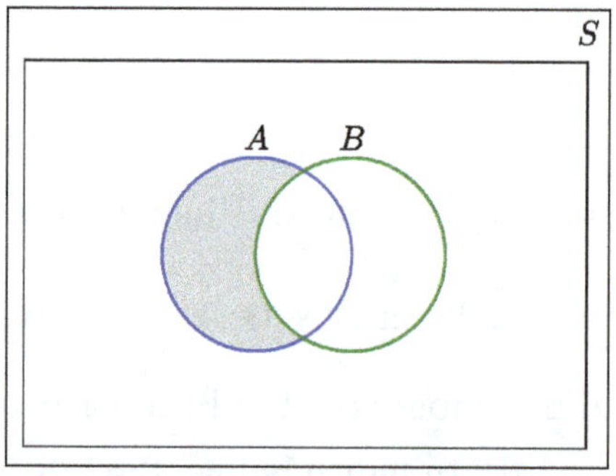

The shaded area shows the set A – B.

Two sets A and B are mutually exclusive or disjoint if they do not have any shared elements; i.e., their intersection is the empty set, $A \cap B = \emptyset$. More generally, several sets are called disjoint if they are pairwise disjoint, i.e., no two of them share a common elements. Figure below shows three disjoint sets.

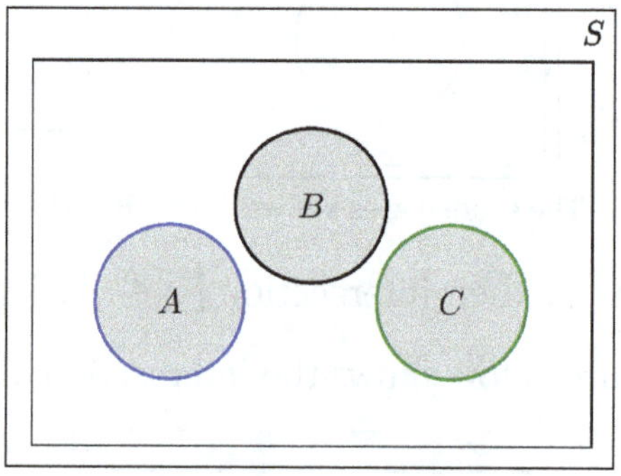

Sets A, B, and C are disjoint.

If the earth's surface is our sample space, we might want to partition it to the different continents. Similarly, a country can be partitioned to different provinces. In general, a collection of nonempty sets A_1, A_2, ... is a partition of a set A if they are disjoint and their union is A. In figure below, the sets A_1, A_2, A_3 and A_4 form a partition of the universal set S.

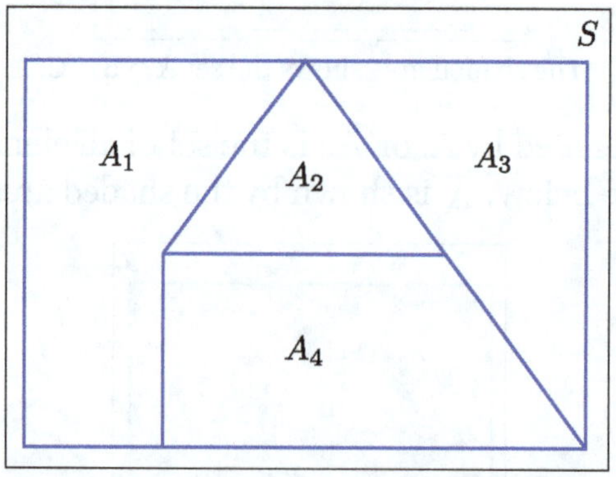

The collection of sets A_1, A_2, A_3 and A_4 is a partition of S.

Here are some rules that are often useful when working with sets. We will see examples of their usage shortly.

Theorem: De Morgan's law. For any sets $A_1, A_2, \cdots, A_n$, we have,

- $(A_1 \cup A_2 \cup A_3 \cup \cdots A_n)^c = A_1^c \cap A_2^c \cap A_3^c \cap A_n^c.$
- $(A_1 \cap A_2 \cap A_3 \cap \cdots A_n)^c = A_1^c \cup A_2^c \cup A_3^c \cup A_n^c.$

Theorem: Distributive law. For any sets A, B, and C we have,

- $A \cap (B \cup C) = (A \cap B) \cup (A \cap C).$
- $A \cup (B \cap C) = (A \cup B) \cap (A \cup C).$

Example: If the universal set is given by S = {1, 2, 3, 4, 5, 6}, and A = {1, 2}, B = {2, 4, 5}, C = {1, 5, 6} are three sets, find the following sets:

- $A \cup B$
- $A \cap B$
- $\bar{A}$
- $\bar{B}$
- Check De Morgan's law by finding $(A \cup B)^c$ and $A^c \cap B^c$.
- Check the distributive law by finding $A \cap (B \cup C)$ and $(A \cap B) \cup (A \cap C)$.

Solution:

- $A \cup B = \{1, 2, 4, 5\}$.
- $A \cap B = \{2\}$.
- $\bar{A} = \{3,4,5,6\}$ ($\bar{A}$ consists of elements that are in S but not in A).
- $\bar{B} = \{1,3,6\}$.
- We have,

  $$(A \cup B)^c = \{1,2,4,5\}^c = \{3,6\},$$

  which is the same as,

  $$A^c \cap B^c = \{3,4,5,6\} \cap \{1,3,6\} = \{3,6\}.$$

- We have,

  $$A \cap (B \cup C) = \{1,2\} \cap \{1,2,4,5,6\} = \{1,2\},$$

  which is the same as,

  $$(A \cap B) \cup (A \cap C) = \{2\} \cup \{1\} = \{1,2\}.$$

A Cartesian product of two sets A and B, written as A × B, is the set containing ordered pairs from A and B. That is, if C = A × B, then each element of C is of the form (x, y), where $x \in A$ and $y \in B$:

$$A \times B = \{(x,y) \mid x \in A \text{ and } y \in B\}.$$

For example, if A = {1, 2, 3} and B = {H, T}, then,

$$A \times B = \{(1,H),(1,T),(2,H),(2,T),(3,H),(3,T)\}.$$

Here the pairs are ordered, so for example, $(1,H) \neq (H,1)$. Thus A × B is not the same as B × A.

If you have two finite sets A and B, where A has M elements and B has N elements, then A × B has M × N elements. This rule is called the multiplication principle and is very useful in counting the numbers of elements in sets. The number of elements in a set is denoted by $|A|$, so here we write $|A| = M, |B| = N$, and $|A \times B| = MN$. In the above example, $|A| = 3$, $|B| = 2$, thus $|A \times B| = 3 \times 2 = 6$. We can similarly define the Cartesian product of n sets $A_1, A_2, \cdots, A_n$ as,

$$A_1 \times A_2 \times A_3 \times \cdots \times A_n = \{(x_1, x_2, \cdots, x_n) \mid x_1 \in A_1 \text{ and } x_2 \in A_2 \text{ and} \cdots x_n \in A_n\}.$$

The multiplication principle states that for finite sets $A_1, A_2, \cdots, A_n$, if,

$$|A_1| = M_1, |A_2| = M_2, \ldots, |A_n| = M_n,$$

then,

$$|A_1 \times A_2 \times A_3 \times \cdots \times A_n| = M_1 \times M_2 \times M_3 \times \cdots \times M_n.$$

An important example of sets obtained using a Cartesian product is R^n, where n is a natural number. For n = 2, we have,

$$\begin{aligned} \mathbb{R}^2 &= \mathbb{R} \times \mathbb{R} \\ &= \{(x,y) \mid x \in \mathbb{R}, y \in \mathbb{R}\}. \end{aligned}$$

Thus, R^2 is the set consisting of all points in the two-dimensional plane. Similarly, $R^3 = \mathbb{R} \times \mathbb{R} \times \mathbb{R}$ and so on.

Finite and Countably Infinite Sets

A set A is said to be finite if either A = ∅ or there is a bijective map $f : A \to \{1, 2, \ldots, n\}$ for some positive integer n (such n is known as the cardinality of the set A). If A = ∅, then A is said to have cardinality 0.

Theorem: Let n be a positive integer. Suppose A is a set and $a_0 \in A$. There exists a bijective map $f: A \to \{1, 2, ..., n+1\}$ if and only if there is a bijective map $g: A \setminus \{a_0\} \to \{1, 2, ..., n\}$.

Proof: Suppose there exists a bijective map $f: A \to \{1, 2, ..., n+1\}$. Suppose further that $f(a_0) = n+1$. Then define $g: A \setminus \{a_0\} \to \{1, 2, ..., n\}$ by $g(x) = f(x)$ for all $x \in A \setminus \{a_0\}$, which is clearly a bijection.

Now let $f(a_0) = m$, for $1 \le m \le n$ and $f(a_1) = n+1$, for some $a_1 \in A$. As f is a bijection, $a_0 \ne a_1$. Define a function $h: A \to \{1, 2, ..., n+1\}$ by $h(x) = f(x)$ for $x \in A \setminus \{a_0, a_1\}$, $h(a_0) = f(a_1)$ and $h(a_1) = f(a_0)$. Then h is also a bijection. Now define $g: A \setminus \{a_0\} \to \{1, 2, ..., n\}$ by $g(x) = h(x)$ for all $x \in A \setminus \{a_0\}$. g is the required bijection. Conversely, let there be a bijection $g: A \setminus \{a_0\} \to \{1, 2, ..., n\}$. Then define $f: A \to \{1, 2, ..., n+1\}$ by $f(x) = g(x)$ for all $x \in A \setminus \{a_0\}$ and $f(a_0) = n + 1$, which serves the purpose.

Theorem: Suppose, $f: A \to \{1, 2, ..., n\}$ is a bijection for some $n \in \mathbb{N}$ and any set A. If B is a proper subset of A then there does not exist any bijection $g: B \to \{1, 2, ..., n\}$. But there exists a bijection $g: B \to \{1, 2, ..., m\}$, for some $m < n$, provided $B \ne \varnothing$.

Proof: If $B = \varnothing$, there cannot exist any bijection $g: B \to \{1, 2, ..., n\}$, for any $n \in \mathbb{N}$. Because, whatever the value of n the set $\{1, 2, ..., n\}$ is nonempty, on the other hand B is empty. Now we apply mathematical induction on n to prove this result.

Suppose n = 1. Then A = {a} and therefore its only proper subset is $B = \varnothing$. Hence the result holds as shown before. Assume that the result holds for some n. We shall establish that the result is true for n + 1. Let $f: A \to \{1,2,...,n+1\}$ be a bijection and B a proper subset of A. If B is empty, the result holds. So, assume that $B \ne \varnothing$. Choose $a_0 \in B$. Since $B \subset A$ and the inclusion is proper, we get $a_1 \in A \setminus B$. There exists a bijection $g: A \setminus \{a_0\} \to \{1,2,...,n\}$. Also, $B \setminus \{a_0\}$ is a proper subset of $A \setminus \{a_0\}$. By induction hypothesis, there does not exist any bijection $h: B \setminus \{a_0\} \to \{1,2,...,n\}$. So, either $B \setminus \{a_0\} = \varnothing$ or else there exists a bijection $k: B \setminus \{a_0\} \to \{1,2,...,p\}$ for some $p < n$. There does not exist a bijection $h^*: B \to \{1,2,...,n+1\}$. Now if $B \setminus \{a_0\} = \varnothing$, then there exists a bijection $k^*: B = \{a_0\} \to \{1\}$. If $B \setminus \{a_0\} \ne \varnothing$, then as there exists a bijection $k: B \setminus \{a_0\} \to \{1,2,...,p\}$, there exists a bijection $k^*: B \to \{1,2,..., p+1\} \to \{1\}$. So in either case, there exists $m < n + 1$, such that there exists a bijection $k^*: B \to \{1,2,...,m\}$, as desired.

Corollary: If A is finite then there is no bijection from A onto a proper subset of A.

Proof: If possible, let there be a bijection $f: A \to B$, where B is a proper subset of A. Since A is a finite set, there exist a bijection $g: A \to \{1,2,...,n\}$, for some $n \in \mathbb{N}$. Consider the map $g \circ f^{-1}: B \to \{1,2,...,n\}$. Since composition of bijective maps is a bijective map, $g \circ f^{-1}$ is a bijection.

Corollary: $\mathbb{N}$ is not finite.

Proof: Consider the map $f:\mathbb{N}\rightarrow\mathbb{N}\setminus\{1\}$, given by $f(n)=n+1$, for each $n\in\mathbb{N}$. It is easy to check that f is a bijection. $\mathbb{N}$ is not finite.

Corollary: The cardinality of a finite set A is uniquely determined by A.

Proof: Let m < n and $f:A\rightarrow\{1,2,\ldots,m\}$, $g:A\rightarrow\{1,2,\ldots,n\}$ be two bijections. Then $g\circ f^{-1}:\{1,2,\ldots,m\}\rightarrow\{1,2,\ldots,n\}$ is also a bijection. Since $\{1,2,\ldots,m\}\subseteq\{1,2,\ldots,n\}$.

Corollary: Any subset of a finite set is finite. If B is a proper subset of a finite set A then the cardinality of B is strictly smaller than the cardinality of A.

Corollary: Let B be a nonempty set. Then the following are equivalent:

- B is finite.
- There is a surjective function from a section of positive integers onto B.
- There is an injective function from B into a section of positive integers.

Proof: (i) ⇒ (ii) Since $B\neq\varnothing$, there is some bijection $f:B\rightarrow\{1,2,\ldots,n\}$, for some $n\in\mathbb{N}$. Hence, f^{-1} serves the requirement. (ii) ⇒ (iii) If $f:\{1,2,\ldots,n\}\rightarrow B$ is surjective then define $g:B\rightarrow\{1,2,\ldots,n\}$ by g(b) = smallest element of $f^{-1}(b)$. g is uniquely defined by the well-ordering property of $\mathbb{N}$ and the fact that f is surjective. To show that g is injective. For any two $b_1\neq b_2$ in B, $f^{-1}(b_1)\cap f^{-1}(b_2)=\varnothing$. So, the smallest element of $f^{-1}(b_1)$ and that of $f^{-1}(b_2)$ are distinct. So, $g(b_1)\neq g(b_2)$. (iii) ⇒ (i) Suppose $f:B\rightarrow\{1,2,\ldots,n\}$ is injective. Consider $\text{Im}(g)\subseteq\{1,2,\ldots,n\}$. Im(g) is finite and hence there is a bijection $g:\text{Im}(g)\rightarrow\{1,2,\ldots,m\}$. Clearly then $g\circ f:B\rightarrow\{1,2,\ldots,m\}$ is a bijection, proving B to be finite.

Corollary: (i) Finite union of finite sets is finite. (ii) Finite Cartesian product of finite sets is finite.

Proof:

1. It is enough to prove the result for union of two sets. Suppose, $A=\varnothing=B$. Then clearly $A\cup B=\varnothing$ and the result holds. If both A, B are non-empty, there exist $f:\{1,2,\ldots,m\}\rightarrow A$ and $g:\{1,2,\ldots,n\}\rightarrow B$ for some m, n. Then define $k:\{1,2,\ldots,m,\ m+1,\ldots m+n\}\rightarrow A\cup B$ by,

$$k(i)=\begin{cases} f(i) & \text{if } i=1,\ 2,\ldots m \\ g(i-m), & \text{if } i=m+1,\ m+2,\ldots\ m+n \end{cases}$$

This map k is clearly bijective and hence, $A\cup B$ is finite.

2. If A or B is empty then $A\times B=\varnothing$. Let A and B be non-empty finite sets. For each $a\in A$, consider $\{a\}\times B$. Then clearly $\{a\}\times B$ is in bijective correspondence with B, by the map $(a,b)\mapsto b$. Hence $\{a\}\times B$ is finite. $A\times B=\bigcup_{a\in A}\{a\}\times B$, i.e., $A\times B$ is a finite union of finite sets and hence by above Corollary, it is finite. Sections of positive integers are prototypes of finite sets. The set of all positive integers constitutes the prototype for a countably infinite set.

A set A is said to be infinite if it is not finite. Moreover, A is countably infinite if there is a bijective map $f : A \to \mathbb{N}$. A set is countable if it is either finite or countably infinite. A set which is not countable is called uncountable.

Example: The set of integers $\mathbb{Z}$ is countably infinite. Define a function $f : \mathbb{Z} \to \mathbb{N}$ by,

$$f(x) = \begin{cases} 2x, & \text{if } x > 0 \\ -2x+1 & \text{otherwise} \end{cases}$$

f is a bijection.

The product $\mathbb{N} \times \mathbb{N}$ is countably infinite. Let $A = \{(x,y) \in \mathbb{N} \times \mathbb{N} : y \le x\}$. Define $f : \mathbb{N} \times \mathbb{N} \to A$ by $f(x,y) = (x+y-1,\ y)$. Then it is easy to check that f is a bijection. Define $g : A \to \mathbb{N}$ by $g(x,y) = \frac{x(x-1)}{2} + y$.

Then g is also a bijection. So, $g \circ f : \mathbb{N} \times \mathbb{N} \to \mathbb{N}$ is a bijection.

Theorem: If C is an infinite subset of $\mathbb{N}$ then C is countably infinite.

Proof: To define a bijection $h : \mathbb{N} \to C$. We proceed by induction as follows: Since C is a non-empty subset of $\mathbb{N}$, C has a smallest element. So, define h(1) = smallest element of C.

Suppose, h(1), ..., h(n – 1) are all defined. Define h(n) = smallest element of $C \setminus h(\{1,2,\dots,n-1\})$. The set $C \setminus h(\{1,2,\dots,n-1\})$ is non-empty and hence, h(n) is well-defined. To show now that h is bijective. It is easy to verify that h is injective. We show that h is surjective.

Let $c \in C$. Since h is injective, $h(\mathbb{N})$ cannot be finite and hence, not contained in {1, 2, ..., c}. So there must be some n such that $h(n) > c$. Suppose, m is the smallest element such that $h(m) \ge c$. So, for all $i < m$, $h(i) < c$. Therefore, $c \notin h(\{1,2,\dots, m-1\})$. i.e., $c \in C \setminus h(\{1,2,\dots, m-1\})$. By definition of h(m), $h(m) \le c$. Hence, h(m) = c.

Theorem: Let B be a nonempty set. The following are equivalent:

- B is countable.
- There is a surjective map $f : \mathbb{N} \to B$.
- There is an injective map $g : B \to \mathbb{N}$.

Proof: (i) ⇒ (ii) Suppose B is countably infinite. Then there is a bijection $h : B \to \mathbb{N}$ and consequently, $f = h^{-1}$ serves the purpose. Let B be a finite set. Then by definition, there exists a bijection $g : B \to \{1,2,\dots,n\}$, for some $n \in \mathbb{N}$. Define a function $f : \mathbb{N} \to B$ by,

$$f(i) = \begin{cases} g(i) & \text{if } 1 \le i \le n \\ g(1), & \text{otherwise} \end{cases}$$

This f is a surjective map. (ii) ⇒ (iii) Define $g : B \to \mathbb{N}$ by g(b) = smallest element of $f^{-1}(b)$. Since f is surjective, each $f^{-1}(b) \neq \varnothing$ and so by applying well-ordering principle g is well-defined. g is injective, because, $g(b_1) = g(b_2) \Rightarrow f^{-1}(b_1) \cap f^{-1}(b_2) \neq \varnothing$ - which is not possible, unless $b_1 = b_2$. (iii) ⇒ (i) Follows from previous theorem.

Corollary: A subset of a countable set is countable.

Proof: Suppose $B \subseteq A$ and A is countable. By previous Theorem (iii), there exists an injective function $f : A \to \mathbb{N}$. Consider $f|_B : B \to \mathbb{N}$. Then $f|B$ is also an injective map and hence by previous theorem again, B is countable.

Corollary: The set $\mathbb{N} \times \mathbb{N}$ is countably infinite.

Proof: In view of the theorem above, it is enough to construct an injective map $f : \mathbb{N} \times \mathbb{N} \to \mathbb{N}$. Define $f : \mathbb{N} \times \mathbb{N} \to \mathbb{N}$ as follows:

$$f(n, m) = 2^n 3^m$$

Suppose, $2^n3^m = 2^p3^q$ for some positive integers n, m, p, q. Let $n < p$. Then $3^m = 2^{p-n}3^m$. But in such case, LHS is a multiple of 3 and RHS is an even number - a contradiction. Similarly, one gets a contradiction if $p < n$. Hence, n = p. So, the equation reduces to $3^m = 3^q$ which of course yields m = q. Therefore, the map f so defined is an injection as desired.

Theorem: A countable union of countable sets is countable.

Proof: Let $\{A_n : n \in J\}$ be a collection of countably many countable sets. So, J is either finite or $\mathbb{N}$. Without loss of generality we assume that each A_n is non-empty. As each A_n is countable, there exists a surjective map $f_n : \mathbb{N} \to A_n$, for each n. Similarly, there exists a surjective map $g : \mathbb{N} \to J$. Define $h : \mathbb{N} \times \mathbb{N} \to \bigcup_{n \in J} A_n$ by $h(k, m) = f_{g(k)}(m)$. Then h is surjective. By previous Corollary, $\mathbb{N} \times \mathbb{N}$ is countably infinite and hence in bijective correspondence with $\mathbb{N}$. Suppose the bijection is $t : \mathbb{N} \to \mathbb{N} \times \mathbb{N}$. So, there exists a surjection $h \circ t : \mathbb{N} \to \bigcup_{n \in J} A_n$ and the result follows from previous theorem.

Theorem: A finite product of countable sets is countable.

Proof: It is enough to show that the product of two countable sets A, B is countable. If A or B = $\varnothing$, then $A \times B = \varnothing$ and so, the result is true. Therefore, without loss of generality we assume that $A \neq \varnothing$ and $B \neq \varnothing$. There are surjective functions $f : \mathbb{N} \to A$ and $g : \mathbb{N} \to B$. Define $h : \mathbb{N} \times \mathbb{N} \to A \times B$ by,

$$h(n, m) = (f(n), g(m))$$

For any $(a, b) \in A \times B$, as f and g are surjective, there exist $n, m \in \mathbb{N}$ such that f(n) = a and g(m) = b. Consequently, (a, b) = (f(n), g(m)) = h(n, m). i.e., h is surjective. On the other hand, there exists a bijection $\eta : \mathbb{N} \to \mathbb{N} \times \mathbb{N}$. Combining these two, we have a surjection $h \circ \eta : \mathbb{N} \to A \times B$, proving that $A \times B$ is countable.

Multisets

A multiset is an unordered collection of elements, in which the multiplicity of an element may be one or more than one or zero. The multiplicity of an element is the number of times the element repeated in the multiset. In other words, we can say that an element can appear any number of times in a set.

Example:

- $A=\{l, l, m, m, n, n, n, n\}$
- $B=\{a, a, a, a, a, c\}$

Operations on Multisets

1. Union of Multisets: The Union of two multisets A and B is a multiset such that the multiplicity of an element is equal to the maximum of the multiplicity of an element in A and B and is denoted by $A \cup B$.

Example:

Let $A=\{l, l, m, m, n, n, n, n\}$, $B=\{l, m, m, m, n\}$,

$A \cup B=\{l, l, m, m, m, n, n, n, n\}$

2. Intersections of Multisets: The intersection of two multisets A and B, is a multiset such that the multiplicity of an element is equal to the minimum of the multiplicity of an element in A and B and is denoted by $A \cap B$.

Example:

Let $A=\{l, l, m, n, p, q, q, r\}$, $B=\{l, m, m, p, q, r, r, r, r\}$

$A \cap B=\{l, m, p, q, r\}$

3. Difference of Multisets: The difference of two multisets A and B, is a multiset such that the multiplicity of an element is equal to the multiplicity of the element in A minus the multiplicity of the element in B if the difference is +ve, and is equal to 0 if the difference is 0 or negative.

Example:

Let $A=\{l, m, m, m, n, n, n, p, p, p\}$, $B=\{l, m, m, m, n, r, r, r\}$

$A-B=\{n, n, p, p, p\}$

4. Sum of Multisets: The sum of two multisets A and B, is a multiset such that the multiplicity of an element is equal to the sum of the multiplicity of an element in A and B.

Example:

$$\text{Let } A=\{l, m, n, p, r\}, \qquad B=\{l, l, m, n, n, n, p, r, r\}$$

$$A+B=\{l, l, l, m, m, n, n, n, n, p, p, r, r, r\}$$

5. Cardinality of Sets: The cardinality of a multiset is the number of distinct elements in a multiset without considering the multiplicity of an element.

Example: $A=\{l, l, m, m, n, n, n, p, p, p, p, q, q, q\}$. The cardinality of the multiset A is 5.

Ordered Set

It is defined as the ordered collection of distinct objects.

Example:

Roll no $\{3, 6, 7, 8, 9\}$
Week Days $\{S, M, T, W, W, TH, F, S, S\}$.

Ordered Pairs

An Ordered Pair consists of two elements such that one of them is designated as the first member and other as the second member. (a, b) and (b, a) are two different ordered pair. An ordered triple can also be written regarding an ordered pair as {(a, b) c}.

An ordered Quadrable is an ordered pair {(((a, b), c) d)} with the first element as ordered triple. An ordered n-tuple is an ordered pair where the first component is an ordered (n - 1) tuples, and the nth element is the second component.

{(n -1), n}

Example: Ordered set of 5 elements:

$$\underbrace{\{(((a,b),c),d)}_{(n-1)}\ \underset{5\text{th}}{\underset{\downarrow}{e}}\}$$

Categories and Free Groups

The language of categories and functors sometimes help to explain complicated situations more efficiently.

A Category C consists of three items:

1. A class of objects, to be denoted by letters A, B, C,...

2. For each pair of objects A and B in $\mathcal{C}$, a set Hom(A, B), called the set of morphisms or arrows $f : A \to B$, $g : A \to B, \ldots$

3. A product Hom(A, B) × Hom(B, C) → Hom(A, C) for each triplet (A, B, C) of objects in $\mathcal{C}$, called composition, denoted by $g \circ f \in \text{Hom}(A, C)$ for f ∈ Hom(A, B) and g ∈ Hom(B, C) such that the following two conditions are satisfied:

- Associativity: For any morphisms $f : A \to B$, $g : B \to C$ and $h : C \to D$,

  $h \circ (g \circ f) = (h \circ g) \circ f$

- (Existence of Identity): For each object A in $\mathcal{C}$, there exists a morphism $I_A : A \to A$, called identity morphism, such that, for all morphisms $f : A \to B$ and $g : C \to A$,

  $f \circ I_A = f$ and $I_A \circ g = g$

Examples of Categories

- Set: The objects are sets and arrows between two sets A and B are the mappings $f : A \to B$.
- Grp: The objects are groups and arrows between two groups G and H are the group-homomorphisms $f : G \to H$.
- Ab: The objects are abelian groups and arrows between two abelian groups G and H are the group-homomorphisms $f : G \to H$.
- Top: The objects are topological spaces and arrows between two topological spaces X and Y are the continuous maps $f : X \to Y$.
- Top∗: The objects are pointed topological spaces and arrows between two pointed topological spaces (X, x) and (Y, y) are the continuous base point preserving maps, i.e. $f : X \to Y$ is continuous with f(x) = y.

The objects in the examples cited above are all of the type 'set with an additional structure' and the morphisms are 'certain functions'. But that is not the case in general. To distinguish this class from the most general ones, these categories are sometimes referred to as 'concrete categories'. So far as our course is concerned, we are interested in concrete categories only. Intuitively, a functor carries the structure of a category to another. i.e., A functor needs to preserve objects as well as morphisms.

Let $\mathcal{C}$ and $\mathcal{D}$ be two categories. A covariant functor from $\mathcal{C}$ to $\mathcal{D}$, denoted by $F : \mathcal{C} \Rightarrow \mathcal{D}$, consists of;

- An object function which assigns to every object A of $\mathcal{C}$ an object F A of $\mathcal{D}$, denoted by $A \mapsto F\,A$.
- An arrow function which assigns to every morphism f ∈ Hom(A, B) of $\mathcal{C}$ a morphism F f ∈ Hom(F A, F B) of $\mathcal{D}$, denoted by $f \mapsto F\,f$; such that,
  - $F(1_A) = 1_{FA}$, for each object A of $\mathcal{C}$ i.e., An identity morphism of $\mathcal{C}$ will be taken to the corresponding identity morphism of $\mathcal{D}$.

- $F(g \circ f) = F g \circ F f$, for any composable pair of morphisms in $\mathcal{C}$. i.e., the commutative diagrams in the category $\mathcal{C}$ will be taken to the commutative diagrams in category $\mathcal{D}$.

Example: (Power set Functor) - Consider the category Set. Define $\mathcal{P} : \text{Set} \Rightarrow \text{Set}$ as follows:

- $\mathcal{P}A$ = Power set of A, for each object A of Set.
- For any $f \in \text{Hom}(A, B)$, $\mathcal{P}(f) \in \text{Hom}(\mathcal{P}A, \mathcal{P}B)$ is such that $\mathcal{P}(f)(S) = f(S)$ whenever $S \subset A$, $f(S) \subset B$.

We observe that $P1_A = 1_{\mathcal{P}A}$ and $\mathcal{P}(g \circ f) = Pg \circ Pf$ hold. So, $\mathcal{P}$ is a functor from Set to Set, called Power set functor.

Example: (Forgetful Functors)

- $F : \text{Grp} \Rightarrow \text{Set}$ is given by,
  - $F(G, *) = G$, where $(G, *)$ is an object of Grp.
  - For any morphism $f : (G, *) \to (H, \circ)$, $F f = f : G \to H$.
- $F : \text{T op} \Rightarrow \text{Set}$ is given by,
  - $F(X, \tau) = X$, where (X, τ) is an object of Top.
  - For any morphism $f : (X, \tau) \to (Y, \sigma)$, $F f = f : X \to Y$.

Let F be an object in a concrete category $\mathcal{C}$, X a non-empty set and $i : X \to F$ a morphism in the category Set. F is said to be free on the set X if for any object A of $\mathcal{C}$ and a morphism $f : X \to A$, there exists a unique morphism of $\mathcal{C}$, $g : F \to A$ such that $g \circ i = f$ in the category Set.

Two objects A and B in a category $\mathcal{C}$ are said to be equivalent (or, isomorphic) if there are morphisms $f : A \to B$ and $g : B \to A$ such that $g \circ f = I_A$ and $f \circ g = I_B$.

Theorem: If $\mathcal{C}$ is a concrete category, F and G are objects of $\mathcal{C}$ such that F is free on the set X and G is free on the set Y and card(X) = card(Y), then F and G are equivalent.

Free Group

Example: Consider the concrete category Grp of groups and homomorphisms. $\mathbb{Z}$ is a free object in Grp. For any group G and $x \in G$, the homomorphism $g : \mathbb{Z} \to G$ given by $g(n) = x^n$ is the unique homomorphism such that $1 \mapsto x$. So, choosing $A = G$, $X = \{1\}$, $i : X \to \mathbb{Z}$ as the inclusion map and $f : X \to G$ given by $f(1) = x$ in the above definition $\mathbb{Z}$ is free on $\{1\}$.

A free object in the concrete category Grp is called a free group. Given a set X, we shall now see the construction of a group F that is free on X.

Case 1: Suppose $X = \varnothing$. Then $F = \{e\}$, the trivial group serves the purpose.

Case 2: Suppose $X \neq \varnothing$. Then we follow the steps as described below:

- Choose a set X^{-1} such that $X \cap X^{-1} = \varnothing$ and $\text{card}(X) = \text{card}(X^{-1})$.

- Since they have the same cardinality, there exists a bijection $f: X \mapsto X^{-1}$. Let $x^{-1} = f(x)$, for each $x \in X$.
- Choose a singleton set A such that $A \cap (X \cup X^{-1}) = \varnothing$. Denote the single element of A by 1.

A word on X is a sequence $(a_1, a_2, \ldots)$ with $a_i \in X \cup X^{-1} \cup \{1\}$ such that $a_k = 1$ for all $k \geq n$, for some $n \in \mathbb{N}$. The constant sequence $(1, 1, \ldots)$, denoted by 1 is called the empty word.

A word is said to be a reduced word, if the following hold:

- For all $x \in X$, x and x^{-1} are not adjacent. i.e., $a_i = x$ implies $a_{i+1} \neq x^{-1}$ and $a_i = x^{-1}$ implies $a_{i+1} \neq x$, for all $i \in \mathbb{N}$.
- $a_k = 1$ implies that $a_i = 1$ for all $i \geq k$.

In particular, empty word is always reduced.

A non-empty reduced word is of the form $\left(x_1^{\alpha_1}, x_2^{\alpha_2}, \ldots, x_n^{\alpha_n}, 1, 1, \ldots\right)$, $n \in \mathbb{N}$, $x_i \in X$ and $\alpha_i = +1$ or -1.

Here, x is identified with x^1. We denote $\left(x_1^{\alpha_1}, x_2^{\alpha_2}, \ldots, x_n^{\alpha_n}, 1, 1, \ldots\right)$ by $x_1^{\alpha_1} x_2^{\alpha_2} \ldots x_n^{\alpha_n}$.

Observation: $x_1^{\alpha_1} x_2^{\alpha_2} \ldots x_n^{\alpha_n} = y_1^{\beta_1} y_2^{\beta_2} \ldots y_m^{\beta_m}$ if and only if one of the following holds:

- Both are empty words.
- $m = n$, $x_i = y_i$ and $\alpha_i = \beta_i$, for all $i = 1, 2, \ldots, n$.

If F(X) is the set of all reduced words on X, then the map $X \to F(X)$, described by $x \mapsto x^1 = x$ is an injective map. We identify X with the copy of X under this map and consider X as a subset of F(X).

Define a binary operation on F(X) following the scheme as itemized below:

$$\left(x_1^{\alpha_1} x_2^{\alpha_2} \ldots x_n^{\alpha_n}\right)\left(y_1^{\beta_1} y_2^{\beta_2} \ldots y_m^{\beta_m}\right) = x_1^{\alpha_1} x_2^{\alpha_2} \ldots x_n^{\alpha_n} y_1^{\beta_1} y_2^{\beta_2} \ldots y_m^{\beta_m}.$$

- The expression on the right is not necessarily a reduced word.
- So, the operation is juxtaposition followed by cancellation of the terms of the form xx^{-1} and $x^{-1}x$. This cancellation is termed as reduction.

It is not hard to check that F(X) with respect to this binary operation of juxtaposition and reduction forms a group.

- The identity element is the empty word 1.
- Inverse of x is x^{-1}, thereby giving $\left(x_1^{-\alpha_1} x_2^{-\alpha_2} \ldots x_n^{-\alpha_n}\right)$ as the inverse of $x_n^{\alpha_n} \ldots x_2^{\alpha_2} \ldots x_1^{\alpha_1}$.

Theorem: F(X) is a free object on the set X, in the concrete category Grp. In view of this theorem, we call F(X) a free group on X and X is called the set of generators of F(X).

Theorem: A subgroup of a free group is free.

A free object on a set X in the category Ab of abelian groups and homomorphisms is called a free abelian group.

Theorem: An abelian group G is a free abelian group if and only if G is isomorphic to an internal direct sum of copies of the additive group $\mathbb{Z}$ of integers.

In particular, the trivial group is a free abelian group on $\varnothing$.

A basis of an abelian group G is a subset S, such that the following hold:

- G is generated by S.
- For distinct $x_1, x_2,..., x_k \in S$ and $n_i \in \mathbb{Z}$,

$$\sum_{i=1}^{k} n_i x_i = 0 \Rightarrow n_1 = n_2 = ... = n_k = 0$$

Theorem: Any two bases of a free abelian group F have the same cardinality.

If X is a basis of a free abelian group F then card(X) is called the rank of F. Finally, the following result establishes a connection between a free group and a free abelian group.

Theorem: A free abelian group is a free group if and only if it is cyclic.

Algebraic Topology

Let X and Y be two topological spaces. The problem, 'Is X homeomorphic to Y' is in general referred to as a Homeomorphism problem. This is a very important and a fundamental question in general topology. Intuitively, two topological spaces are considered same, if they are homeomorphic.

Example: An open interval (i.e., a line segment without its end points) and a semicircular arc (without its end points) are homeomorphic. An open interval is nothing but a line segment without its end points, whereas a semi-circular arc (without end-points) is a bended line segment. So the obvious way of getting the homeomorphism is just flattening the circular arc into a straight line. The actual correspondence is obtained trivially as follows:

- Draw a line through the centre of the semi-circular arc that intersects both the objects.
- This line cuts the semicircle at x and the open interval at y.
- The correspondence is $x \leftrightarrow y$.

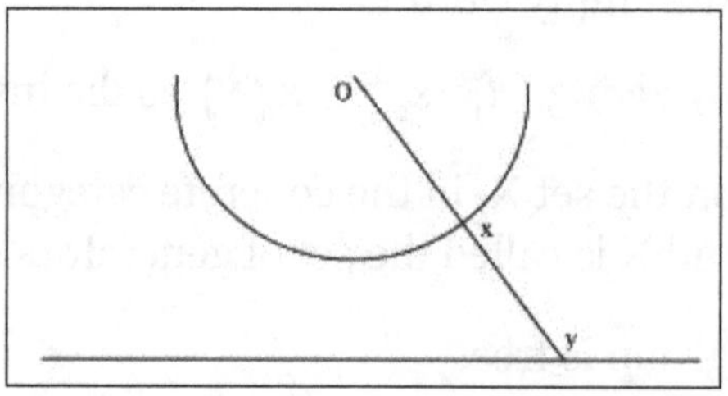

Another well-known technique in point set topology, called puncturing objects, is also used to establish that two spaces are not topologically same (i.e., not homeomorphic).

Example: A circle is not homeomorphic to an open interval. Removing a point from the circle (this removal is called puncturing) transforms it to a single line segment. On the other hand, removing a point from the open interval breaks the open interval into two pieces of line segments. A single line segment (i.e., a space with only one component) can't be homeomorphic to two pieces of lines (i.e., a space with two components), and consequently, the original figures are not homeomorphic. (Here, the argument depends on the fact that connectedness is a topological property and so, a connected space cannot be homeomorphic to a disconnected space).

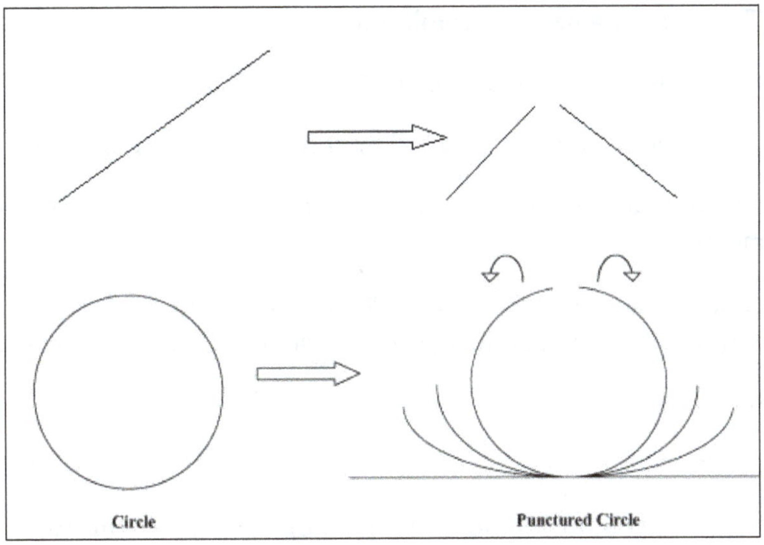

Example: The real line $\mathbb{R}$ is not homeomorphic to R^2. Removal of one point from $\mathbb{R}$ breaks the real line into two components (hence, punctured $\mathbb{R}$ is disconnected), but R^2 is still connected when a single point is removed from it. A connected space cannot be homeomorphic to a disconnected space (as connectedness is a topological property) and so, the original spaces are not homeomorphic to each other.

In fact, in the last example, there is nothing special about R^2 and so, the same arguments establish that R^n is not homeomorphic to $\mathbb{R}$, for any $n > 1$. But, it is not possible to argue in the same line to obtain in general that R^m is not homeomorphic to R^n, for any $m \neq n$.

Example: In R^3, a sphere (S^2) is not homoemorphic to a torus ($S^1 \times S^1$). If we remove a circle from the torus T (as indicated in the following figure), we get a pipelike shape. This new shape remains connected. But removing a circle from a sphere S^2 (as shown in the next diagram) breaks the sphere into two parts (i.e., the result is a disconnected space). Hence, a torus is not homeomorphic to a sphere in Euclidean 3-space.

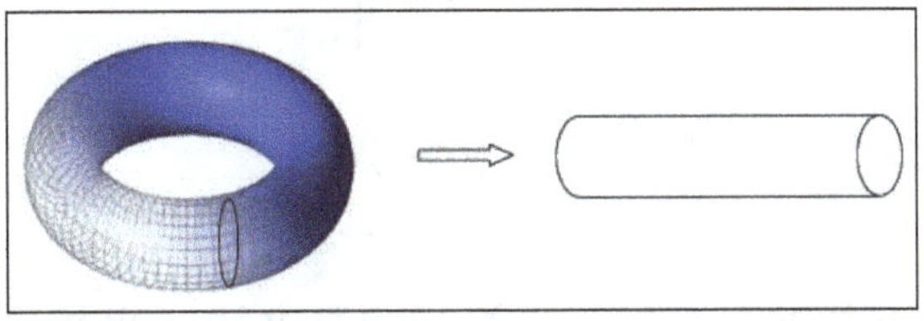

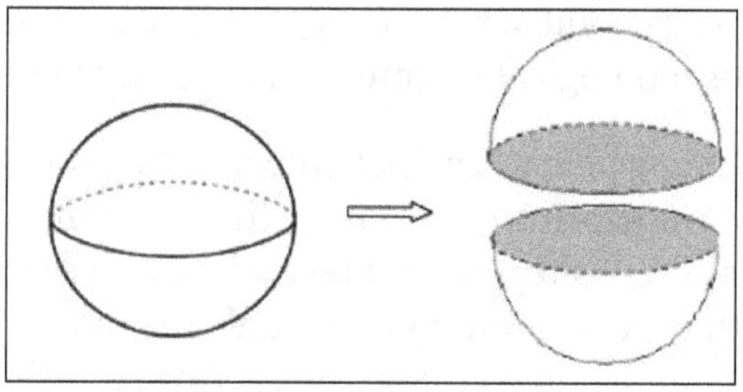

Unlike the case of a 2-sphere S^2 and a torus (i.e., $S^1 \times S^1$), it is highly difficult even to prove that S^3 is not homeomorphic to $S^1 \times S^1 \times S^1$. Algebraic topology partially solves this so-called 'Homeomorphism problems '. The general scheme is as follows:

- To each topological space X, associate an algebraic object H(X).
- If there is a continuous function $X \to Y$ then it induces a homomorphism $H(X) \to H(Y)$.
- If the induced homomorphism is not an isomorphism then the given topological spaces are not homeomorphic.

The algebraic object H(X) may be a group, module, a sequence of modules, etc. There are several methods of construction of such algebraic objects. We shall explore two constructions of H(X):

- Fundamental group/Homotopy group.
- Homology group.

By using the technique of algebraic topology, homeomorphism problems can be solved partially - not completely. In fact,

$$X \xrightarrow{\text{homeo}} Y \Rightarrow H(X) \xrightarrow{\text{iso}} H(Y)$$

But the converse of this result is not in general true. So, this result helps for establishing not same rather than sameness of two spaces.

However, there are situations where this implication is both-sided. For example, consider the following: Suppose $p : \tilde{X} \to X$ is an onto continuous map. If $f : Y \to X$ is a given continuous function then 'Does there exist a continuous function $g : Y \to \tilde{X}$ such that $p \circ g = f$?' This is known as Lifting problem in set-topology. If such a map g exists then g is called a lift of f. We restate this problem in terms of commutativity of diagrams: Suppose,

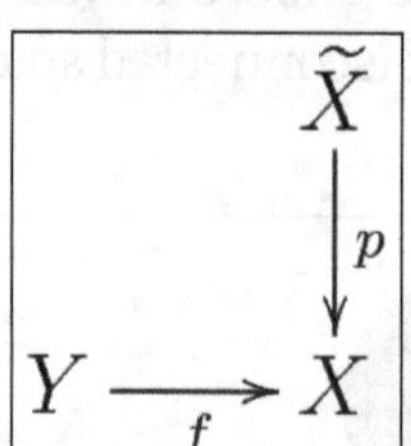

Does there exist a continuous function $g : Y \to \tilde{X}$ such that the following diagram commutes?

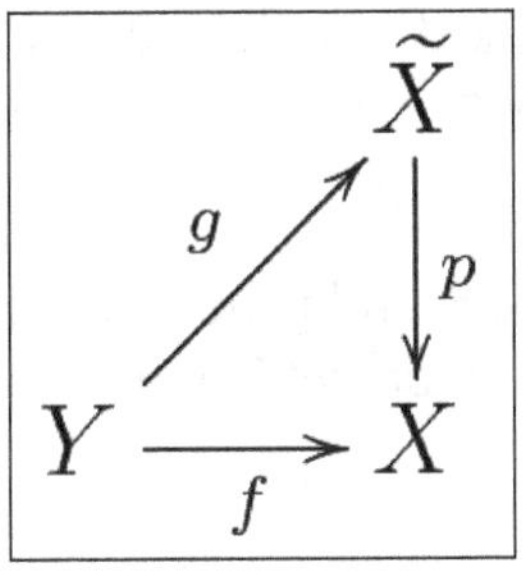

It can be established that the commutativity of the diagram of groups and group homomorphisms completely determines the commutativity of the diagram of topological spaces and continuous maps:

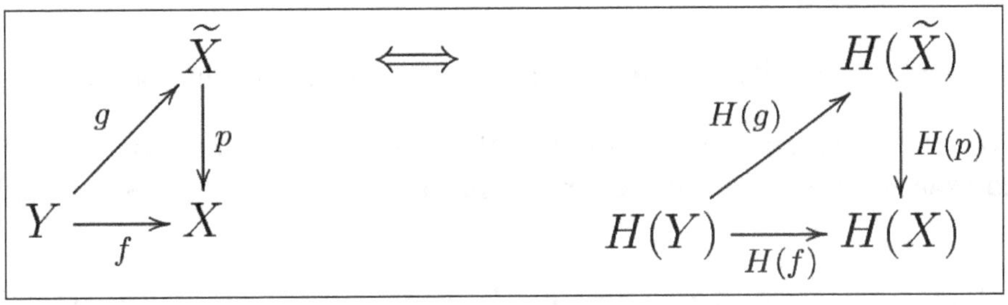

So, in a nutshell, Algebraic topology aims at creating and defining various algebraic objects associated to topological spaces and homomorphisms induced by continuous maps. We observe that the algebraic objects so associated depend on the topology of X in such a manner that if $f : X \to Y$ is a continuous map between X and Y, then it naturally induces a homomorphism H(f) between the groups H(X) and H(Y). Furthermore, the induced homomorphisms have the following two properties:

- If $f : X \to Y$, $g : Y \to Z$ are continuous maps, then the induced homomorphisms H(f), H(g) and $H(g \circ f)$ are related by,

  $$H(g \circ f) = H(g) \circ H(f)$$

- If $I_X : X \to X$ is the identity map of the space X, then $H(I_X) = I_{H(X)}$ is the identity map of the group H(X).

It is evident that these properties establish that H is a functor from the Category of topological spaces to the category of groups. These two properties are thus referred to as functorial properties.

Inclusion-Exclusion Principle in Set Theory

The principle of inclusion and exclusion (PIE) is a counting technique that computes the number of elements that satisfy at least one of several properties while guaranteeing that elements satisfying more than one property are not counted twice.

An underlying idea behind PIE is that summing the number of elements that satisfy at least one of two categories and subtracting the overlap prevents double counting. For instance, the number of people that have at least one cat or at least one dog can be found by taking the number of people who own a cat, adding the number of people that have a dog, then subtracting the number of people who have both.

PIE is particularly useful in combinatorics and probability problem solving when it is necessary to devise a counting method that ensures an object is not counted twice.

Two Sets

In the case of objects being separated into two (possibly disjoint) sets, the principle of inclusion and exclusion states,

$$|A \cup B| = |A| + |B| - |A \cap B|,$$

where $|S|$ denotes the cardinality, or number of elements, of set S in set notation.

Proof: To prove this statement, we will show that every element which belongs in one of these sets is counted exactly once, and every element that is not in these sets is counted exactly zero times.

- Case 1: Element is not in A, and not in B. It is obvious it is counted zero times in the LHS. It is obvious it is counted zero times in the RHS.
- Case 2: Element is in A, and not in B. It will be counted once on the LHS. On the RHS, it is counted once in $|A|$.
- Case 3: Element is not in A, and is in B. It will be counted once on the LHS. On the RHS, it is counted once in $|B|$.
- Case 4: Element is in A, and in B. It will be counted once on the LHS. On the RHS, it is counted +1 in $|A|$,+1 in $|B|$ and −1 in $|A \cap B|$. Hence, it is counted exactly once.

As a Venn diagram, PIE for two sets can be depicted easily:

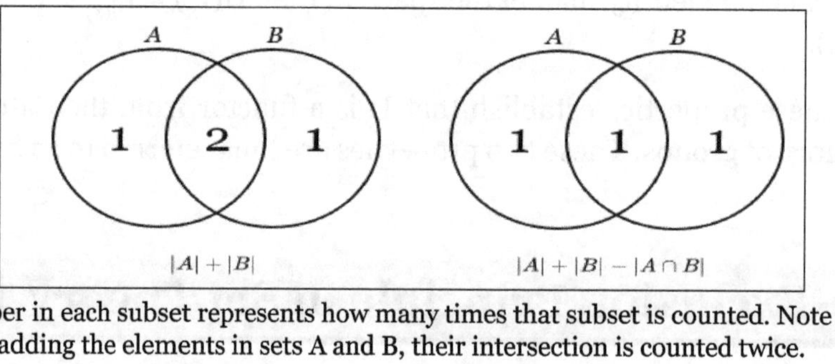

The number in each subset represents how many times that subset is counted. Note that by adding the elements in sets A and B, their intersection is counted twice.

Example: How many integers from 1 to 100 are multiples of 2 or 3? Let A be the set of integers from 1 to 100 that are multiples of 2, then $|A| = 50$. Let B be the set of integers from 1 to 100 that are multiples of 3, then $|B| = 33$. Now, $A \cap B$ is the set of integers from 1 to 100 that are multiples of both 2 and 3, and hence are multiples of 6, implying $|A \cap B| = 16$.

Hence, by PIE,

$$|A \cup B| = |A| + |B| - |A \cap B| = 50 + 33 - 16 = 67.$$

Three Sets and More

We've already looked at the case of 2 sets. Before we dive into the general case, let's consider having 3 sets. If there are three sets, the principle of inclusion and exclusion states,

$$|A \cup B \cup C| = |A| + |B| + |C| - |A \cap B| - |A \cap C| - |B \cap C| + |A \cap B \cap C|.$$

We can verify these statements for ourselves by considering the Venn diagram of events:

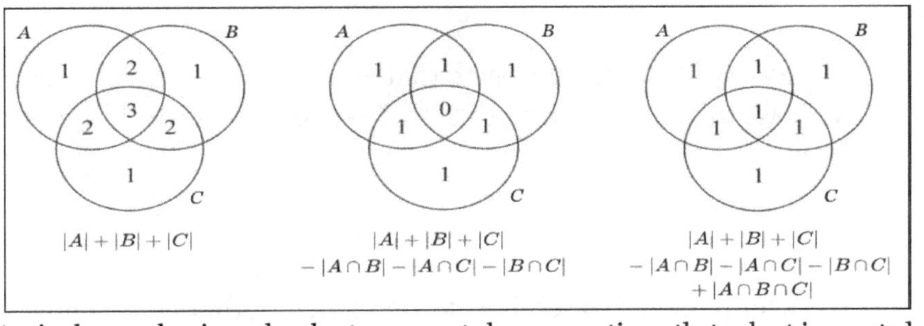

Again the number in each subset represents how many times that subset is counted. Here, the overlap of all three subsets must be added back in due to it being removed after the second step.

Example: There are exactly three types of students in a school: the geeks, the wannabees, and the athletes. Each student is classified into at least one of these categories. And the total number of students in the school is 1000. Suppose that the following is given:

- The total number of students who are geeks is 310.
- The total number of students who are wannabees is 650.
- The total number of students who are athletes is 440.
- The total number of students who are both geeks and wannabees is 170.
- The total number of students who are both geeks and athletes is 150.
- The total number of students who are both wannabees and athletes is 180.

What is the total number of students who fit into all 3 categories?

Let G, W, A denote the set for geeks, wannabees, and athletes, respectively. Then by the principle of inclusion and exclusion, we have,

$$n(G \cup W \cup A) = n(G) + n(W) + n(A) - n(W \cap G) - n(G \cap A) - n(W \cap A) + n(G \cap W \cap A),$$

which gives us $1000 = 310 + 650 + 440 - 170 - 150 - 180 + n(G \cap W \cap A) = 900 + n(G \cap W \cap A)$.

So the total number of students who fit into all 3 categories is 100.

You can duplicate the proof above and check that every element in $A \cup B \cup C$ is counted exactly once on the RHS.

More generally, if $A_1, A_2, \ldots, A_n$ are finite sets, then the principle of inclusion and exclusion states,

$$\left|\bigcup_{i=a}^{n} A_i\right| = \sum_{i=1}^{n}|A_i| - \sum_{1 \le i < j \le n}|A_i \cap A_j| + \sum_{1 \le i < j < k \le n}|A_i \cap A_j \cap A_k| - \cdots + (-1)^{n-1}|A_i \cap \cdots \cap A_n|$$

Derangements

There are eight guests at a secret Santa party. Each guest brings a gift and each receives another gift in return. No one is allowed to receive the gift they brought. How many ways are there to distribute the gifts?

Let A denote the set of ways to distribute gifts such that everyone receives a gift, possibly their own. Let A_i denote the set of ways to distribute gifts such that person ii receives his or her own gift. Then we would like to find,

$$|A| - |A_1 \cup A_2 \cup \cdots \cup A_8|$$

Since A_i is the set of ways for person ii to receive his/her own gift, there are 7 choices of gifts for the next person, 6 choices of gifts for the following person, and so on. By the rule of product,

$$|A_i| = 7 \times 6 \times \cdots \times 2 \times 1 = 7!.$$

Since $A_i \cap A_j$ is the set of ways person ii and person j both receive their own gifts, there are 6 choices of gifts for the next person, 5 choices of gifts for the following person, and so on. Again by the rule of product,

$$|A_i \cap A_j| = 6 \times 5 \times \cdots \times 2 \times 1 = 6!$$

By continuing this argument, if k people receive their own gifts, then there are (8 - k)! possible ways. Applying PIE, we obtain,

$$|A| - |A_1 \cup A_2 \cup \cdots \cup A_8| = 8! - \binom{8}{1} \times 7! + \binom{8}{2} \times 6! - \cdots + \binom{8}{8} \times 1! = 14833$$

A derangement of n objects is a permutation of the objects such that none of them stays in the same place. The number of ways this can be done is denoted D_n, and this calculation shows $D_8 = 14833$.

Example: What is the sum of all integers from 1 to 100 that are multiples of 2 or 3? While PIE is often used to count the elements of a set, if we remove the $|\cdot|$ symbols, the statement is still true. For example, in two variables, $A \cup B = A + B - A \cap B$. The same proof using Venn diagrams works to show that each element is included once. As such, the sum of elements in $A \cup B$ is equal to the sum of elements in A plus the sum of elements in B minus the sum of elements in $A \cap B$. Let A be the set of multiples of 2 and B be the set of multiples of 3, then $A \cap B$ is the set of multiples of 6, and hence the sum $A \cup B$ is,

$$\frac{(2+100)\times 50}{2}+\frac{(3+99)\times 33}{2}-\frac{(6+96)\times 16}{2}=3417$$

Example: We have 7 balls each of different colors (red, orange, yellow, green, blue, indigo, violet) and 3 boxes each of different shapes (tetrahedron, cube, dodecahedron). How many ways are there to place these 7 balls into the 3 boxes such that each box contains at least 1 ball?

Let X be the total number of ways we can distribute the balls if there are no restrictions. Each ball can be placed into any one of the 3 boxes, so $|X| = 3^7$. Let T be the set of ways such that the tetrahedron box has no balls, C the set of ways such that the cube box has no balls, and D the set of ways such that the dodecahedron box has no balls. We would like to find,

$$|X| - |T \cup C \cup D|$$

We have $|T| = |C| = |D| = 2^7$, since the balls can be placed into one of the two other boxes, and $|T \cap C| = |C \cap D| = |D \cap T| = 1^7$, since all the balls must be placed in the remaining box, and $|T \cap C \cap D| = 0$. Hence, PIE implies,

$$|X| - |T \cup C \cup D| = 3^7 - 3 \times 2^7 + 3 \times 1^7 - 0 = 1806$$

Relations

Relation or Binary relation R from set A to B is a subset of AxB which can be defined as,

$$aRb \leftrightarrow (a,b) € R \leftrightarrow R(a,b)$$

A Binary relation R on a single set A is defined as a subset of AxA. For two distinct set, A and B with cardinalities m and n, the maximum cardinality of the relation R from A to B is mn.

Domain and Range: If there are two sets A and B and Relation from A to B is R(a,b), then domain is defined as the set {a | (a,b) € R for some b in B} and Range is defined as the set {b | (a,b) € R for some a in A}.

Types of Relation

- Empty Relation: A relation R on a set A is called Empty if the set A is empty set.
- Full Relation: A binary relation R on a set A and B is called full if AXB.
- Reflexive Relation: A relation R on a set A is called reflexive if (a,a) € R holds for every element a € A i.e. if set A = {a,b} then R = {(a,a), (b,b)} is reflexive relation.
- Irreflexive relation: A relation R on a set A is called reflexive if no (a,a) € R holds for every element a € A i.e. if set A = {a,b} then R = {(a,b), (b,a)} is irreflexive relation.

- Symmetric Relation: A relation R on a set A is called symmetric if (b,a) € R holds when (a,b) € R i.e. the relation R = {(4,5), (5,4), (6,5), (5,6)} on set A = {4,5,6} is symmetric.
- Antisymmetric Relation: A relation R on a set A is called antisymmetric if (a,b) € R and (b,a) € R then a = b is called antisymmetric. i.e. the relation R = $\{(a,b) \rightarrow R | a \leq b\}$ is anti-symmetric since $a \leq b$ and $b \leq a$ implies a = b.
- Transitive Relation: A relation R on a set A is called transitive if (a,b) € R and (b,c) € R then (a,c) € R for all a,b,c € A i.e. Relation R = {(1,2), (2,3), (1,3)} on set A = {1,2,3} is transitive.
- Equivalence Relation: A relation is an Equivalence Relation if it is reflexive, symmetric, and transitive. i.e. relation R = {(1,1), (2,2), (3,3), (1,2), (2,1), (2,3), (3,2), (1,3), (3,1)} on set A = {1,2,3} is equivalence relation as it is reflexive, symmetric, and transitive.
- Asymmetric relation: Asymmetric relation is opposite of symmetric relation. A relation R on a set A is called asymmetric if no (b,a) € R when (a,b) € R.

Important Points:

1. Symmetric and anti-symmetric relations are not opposite because a relation R can contain both the properties or may not.

2. A relation is asymmetric if and only if it is both anti-symmetric and irreflexive.

3. Number of different relation from a set with n elements to a set with m elements is 2^{mn}.

Example: if $R = \{r_{a1}, r_2, r_3 \ldots\ldots r_n\}$ and $S = \{s_1, s_2, s_3 \ldots\ldots s_m\}$

then Cartesian product of R and S is:

$$\begin{aligned} R \text{ X } S = \{&(r_1, s_1), (r_1, s_2), (r_1, s_3)\ldots\ldots\ldots, (r_1, s_n), \\ &(r_2, s_1), (r_2, s_2), (r_2, s_3)\ldots\ldots\ldots\ldots, (r_2, s_n), \\ &\ldots\ldots\ldots\ldots\ldots\ldots \\ &(r_n, s_1),(r_n, s_2), (r_n, s_3),\ldots\ldots\ldots, (r_n, s_n)\} \end{aligned}$$

This set of ordered pairs contains mn pairs. Now these pairs can be present in R X S or can be absent. So total number of possible relation = 2^{mn}.

4. Number of Reflexive Relations on a set with n elements: $2^{n(n-1)}$: A relation has ordered pairs (a,b). Now a can be chosen in n ways and same for b. So set of ordered pairs contains n^2 pairs. Now for a reflexive relation, (a,a) must be present in these ordered pairs. And there will be total n pairs of (a,a), so number of ordered pairs will be n^2-n pairs. So total number of reflexive relations is equal to $2^{n(n-1)}$.

5. Number of Symmetric Relations on a set with n elements: $2^{n(n+1)/2}$: A relation has ordered pairs (a,b). Now for a symmetric relation, if (a,b) is present in R, then (b,a) must be present in R. In Matrix form, if a_{12} is present in relation, then a_{21} is also present in relation and as we know reflexive

relation is part of symmetric relation. So from total n^2 pairs, only n(n+1)/2 pairs will be chosen for symmetric relation. So total number of symmetric relation will be $2^{n(n+1)/2}$.

6. Number of Anti-Symmetric Relations on a set with n elements: $2^n\ 3^{n(n-1)/2}$. A relation has ordered pairs (a,b). For anti-symmetric relation, if (a,b) and (b,a) is present in relation R, then a = b.(That means a is in relation with itself for any a). So for (a,a), total number of ordered pairs = n and total number of relation = 2^n.

If (a,b) and (b,a) both are not present in relation or Either (a,b) or (b,a) is not present in relation. So there are three possibilities and total number of ordered pairs for this condition is n(n-1)/2. (Selecting a pair is same as selecting the two numbers from n without repetition). As we have to find number of ordered pairs where a ≠ b. it is like opposite of symmetric relation means total number of ordered pairs = (n^2) – symmetric ordered pairs(n(n+1)/2) = n(n-1)/2. So, total number of relation is $3^{n(n-1)/2}$. So total number of anti-symmetric relation is $2^n.3^{n(n-1)/2}$.

7. Number of Asymmetric Relations on a set with n elements: $3^{n(n-1)/2}$. In Asymmetric Relations, element a cannot be in relation with itself. (i.e. there is no aRa ∀ a∈A relation.) And then it is same as Anti-Symmetric Relations. (i.e. you have three choice for pairs (a,b) (b,a)). Therefore there are $3^{n(n-1)/2}$ Asymmetric Relations possible.

8. Irreflexive Relations on a set with n elements: $2^{n(n-1)}$. A relation has ordered pairs (a,b). For Irreflexive relation, no (a,a) holds for every element a in R. It is also opposite of reflexive relation. Now for a Irreflexive relation, (a,a) must not be present in these ordered pairs means total n pairs of (a,a) is not present in R, So number of ordered pairs will be n^2-n pairs. So total number of reflexive relations is equal to $2^{n(n-1)}$.

9. Reflexive and symmetric Relations on a set with n elements: $2^{n(n-1)/2}$. A relation has ordered pairs (a,b). Reflexive and symmetric Relations means (a,a) is included in R and (a,b)(b,a) pairs can be included or not. (In Symmetric relation for pair (a,b)(b,a) (considered as a pair). whether it is included in relation or not) So total number of Reflexive and symmetric Relations is $2^{n(n-1)/2}$.

Functions

A Function assigns to each element of a set, exactly one element of a related set. Functions find their application in various fields like representation of the computational complexity of algorithms, counting objects, study of sequences and strings, to name a few.

A function or mapping (Defined as $f : X \rightarrow Y$) is a relationship from elements of one set X to elements of another set Y (X and Y are non-empty sets). X is called Domain and Y is called Codomain of function 'f'.

Function 'f' is a relation on X and Y such that for each x ∈ X, there exists a unique y ∈ Y such that (x,y) ∈ R. 'x' is called pre-image and 'y' is called image of function f. A function can be one to one or many to one but not one to many.

Injective/One-to-one Function

A function $f: A \to B$ is injective or one-to-one function if for every $b \in B$, there exists at most one $a \in A$ such that f(s) = t. This means a function f is injective if $a_1 \neq a_2$ implies f(a1) ≠ f(a2).

Example:

- $f: N \to N, f(x) = 5x$ is injective.
- $f: N \to N, f(x) = x^2$ is injective.
- $f: R \to R, f(x) = x^2$ is not injective as $(-x)^2 = x^2$.

Surjective/Onto Function

A function $f: A \to B$ is surjective (onto) if the image of f equals its range. Equivalently, for every $b \in B$, there exists some $a \in A$ such that f(a) = b. This means that for any y in B, there exists some x in A such that y = f(x).

Example:

- $f: N \to N, f(x) = x + 2$ is surjective.
- $f: R \to R, f(x) = x^2$ is not surjective since we cannot find a real number whose square is negative.

Bijective/One-to-one Correspondent

A function $f: A \to B$ is bijective or one-to-one correspondent if and only if f is both injective and surjective.

Problem: Prove that a function $f: R \to R$ defined by f(x) = 2x – 3 is a bijective function.

solution: We have to prove this function is both injective and surjective. If $f(x_1) = f(x_2)$, then $2x_1 - 3 = 2x_2 - 3$ and it implies that $x_1 = x_2$.

Hence, f is injective.

Here, $2x - 3 = y$

So, $x = (y+5)/3$ which belongs to R and $f(x) = y$.

Hence, f is surjective. Since f is both surjective and injective, we can say f is bijective.

Inverse of a Function

The inverse of a one-to-one corresponding function $f: A \to B$, is the function $g: B \to A$, holding the following property:

$$f(x) = y \Leftrightarrow g(y) = x$$

The function f is called invertible, if its inverse function g exists.

Example:

- A Function $f:Z \to Z, f(x)=x+5$, is invertible since it has the inverse function $g:Z \to Z,\ g(x)=x-5$.
- A Function $f:Z \to Z, f(x)=x^2$ is not invertiable since this is not one-to-one as $(-x)^2=x^2$.

Composition of Functions

Two functions $f: A \to B$ and $g: B \to C$ can be composed to give a composition gof. This is a function from A to C defined by $(gof)(x)=g(f(x))$.

Example: Let $f(x)=x+2$ and $g(x)=2x+1$, find (fog)(x) and (gof)(x).

Solution:

$$(fog)(x)=f(g(x))=f(2x+1)=2x+1+2=2x+3$$

$$(gof)(x)=g(f(x))=g(x+2)=2(x+2)+1=2x+5$$

Hence, $(fog)(x) \neq (gof)(x)$.

Some Facts about Composition

- If f and g are one-to-one then the function (gof) is also one-to-one.
- If f and g are onto then the function (gof) is also onto.
- Composition always holds associative property but does not hold commutative property.

References

- Discrete-mathematics-sets: tutorialspoint.com, Retrieved 11, April 2020
- Algebra-of-sets: javatpoint.com, Retrieved 24, June 2020
- Multisets: javatpoint.com, Retrieved 06, March 2020
- Principle-of-inclusion-and-exclusion-pie: brilliant.org, Retrieved 17, January 2020
- Relations-and-their-types: geeksforgeeks.org, Retrieved 30, August 2020
- Discrete-mathematics-functions: tutorialspoint.com, Retrieved 09, May 2020

4

Number Theory and its Types

The branch of mathematics which deals with integers and integer value functions is called number theory. A few of its major branches are analytic number theory and algebraic number theory. This chapter closely examines these facets of number theory to provide an extensive understanding of the subject.

Number theory or, in older usage, arithmetic is a branch of pure mathematics devoted primarily to the study of the integers. It is sometimes called "The Queen of Mathematics" because of its foundational place in the discipline. Number theorists study prime numbers as well as the properties of objects made out of integers (e.g., rational numbers) or defined as generalizations of the integers (e.g., algebraic integers).

A Lehmer sieve, which is a primitive digital computer once used for finding primes and solving simple Diophantine equations.

Integers can be considered either in themselves or as solutions to equations (Diophantine geometry). Questions in number theory are often best understood through the study of analytical objects (e.g., the Riemann zeta function) that encode properties of the integers, primes or other number-theoretic objects in some fashion (analytic number theory). One may also study real numbers in relation to rational numbers, e.g., as approximated by the latter (Diophantine approximation).

The older term for number theory is *arithmetic*. By the early twentieth century, it had been superseded by "number theory". (The word "arithmetic" is used by the general public to mean "elementary calculations"; it has also acquired other meanings in mathematical logic, as in *Peano arithmetic*, and computer science, as in *floating point arithmetic*.) The use of the term *arithmetic* for *number theory* regained some ground in the second half of the 20th century, arguably in part due to French influence. In particular, *arithmetical* is preferred as an adjective to *number-theoretic*.

History

Dawn of Arithmetic

The first historical find of an arithmetical nature is a fragment of a table: the broken clay tablet Plimpton 322 (Larsa, Mesopotamia, ca. 1800 BCE) contains a list of "Pythagorean triples", i.e., integers (a,b,c) such that $a^2+b^2=c^2$. The triples are too many and too large to have been obtained by brute force. The heading over the first column reads: "The *takiltum* of the diagonal which has been subtracted such that the width..."

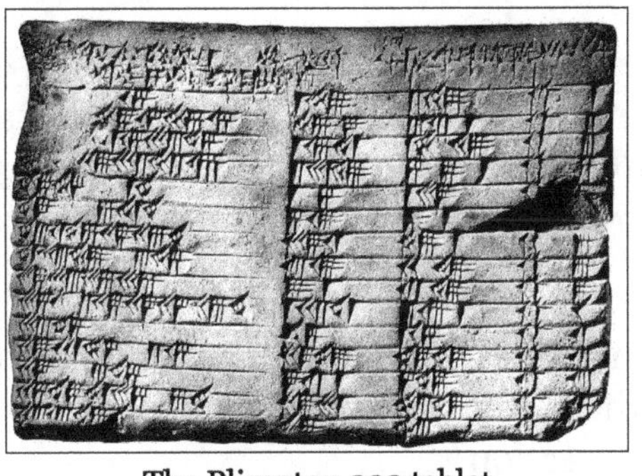

The Plimpton 322 tablet.

The table's layout suggests that it was constructed by means of what amounts, in modern language, to the identity

$$\left(\frac{1}{2}\left(x-\frac{1}{x}\right)\right)^2+1=\left(\frac{1}{2}\left(x+\frac{1}{x}\right)\right)^2,$$

which is implicit in routine Old Babylonian exercises. If some other method was used, the triples were first constructed and then reordered by $c>$, presumably for actual use as a "table", i.e., with a view to applications.

It is not known what these applications may have been, or whether there could have been any; Babylonian astronomy, for example, truly flowered only later. It has been suggested instead that the table was a source of numerical examples for school problems.

While Babylonian number theory—or what survives of Babylonian mathematics that can be called thus—consists of this single, striking fragment, Babylonian algebra (in the secondary-school sense of "algebra") was exceptionally well developed. Late Neoplatonic sources state that Pythagoras learned mathematics from the Babylonians. Much earlier sources state that Thales and Pythagoras traveled and studied in Egypt.

Euclid IX 21—34 is very probably Pythagorean; it is very simple material ("odd times even is even", "if an odd number measures [= divides] an even number, then it also measures [= divides] half of it"), but it is all that is needed to prove that $\sqrt{2}$ is irrational. Pythagorean mystics gave great importance to the odd and the even. The discovery that $\sqrt{2}$ is irrational is credited to the early Pythagoreans (pre-Theodorus). By revealing (in modern terms) that numbers could be irrational, this discovery seems to have provoked the first foundational crisis in mathematical history; its proof or its divulgation are sometimes credited to Hippasus, who was expelled or split from the Pythagorean sect. This forced a distinction between *numbers* (integers and the rationals—the subjects of arithmetic), on the one hand, and *lengths* and *proportions* (which we would identify with real numbers, whether rational or not), on the other hand.

The Pythagorean tradition spoke also of so-called polygonal or figurate numbers. While square numbers, cubic numbers, etc., are seen now as more natural than triangular numbers, pentagonal numbers, etc., the study of the sums of triangular and pentagonal numbers would prove fruitful in the early modern period (17th to early 19th century).

We know of no clearly arithmetical material in ancient Egyptian or Vedic sources, though there is some algebra in both. The Chinese remainder theorem appears as an exercise in Sun Zi's *Suan Ching*, also known as *The Mathematical Classic of Sun Zi* (3rd, 4th or 5th century CE.) (There is one important step glossed over in Sun Zi's solution: it is the problem that was later solved by Āryabhaṭa's Kuṭṭaka.)

There is also some numerical mysticism in Chinese mathematics, but, unlike that of the Pythagoreans, it seems to have led nowhere. Like the Pythagoreans' perfect numbers, magic squares have passed from superstition into recreation.

Classical Greece and the Early Hellenistic Period

Aside from a few fragments, the mathematics of Classical Greece is known to us either through the reports of contemporary non-mathematicians or through mathematical works from the early Hellenistic period. In the case of number theory, this means, by and large, *Plato* and *Euclid*, respectively.

Plato had a keen interest in mathematics, and distinguished clearly between arithmetic and calculation. (By *arithmetic* he meant, in part, theorising on number, rather than what *arithmetic* or *number theory* have come to mean.) It is through one of Plato's dialogues—namely, *Theaetetus*—that we know that Theodorus had proven that $\sqrt{3}, \sqrt{5}, \ldots, \sqrt{17}$ are irrational. Theaetetus was, like Plato, a disciple of Theodorus's; he worked on distinguishing different kinds of incommensurables, and was thus arguably a pioneer in the study of number systems. (Book X of Euclid's Elements is described by Pappus as being largely based on Theaetetus's work.)

Euclid devoted part of his *Elements* to prime numbers and divisibility, topics that belong unambiguously to number theory and are basic to it (Books VII to IX of Euclid's Elements). In particular, he gave an algorithm for computing the greatest common divisor of two numbers (the Euclidean algorithm; *Elements*, Prop. VII.2) and the first known proof of the infinitude of primes (*Elements*, Prop. IX.20).

In 1773, Lessing published an epigram he had found in a manuscript during his work as a librarian; it claimed to be a letter sent by Archimedes to Eratosthenes. The epigram proposed what has become known as Archimedes' cattle problem; its solution (absent from the manuscript) requires solving an indeterminate quadratic equation (which reduces to what would later be misnamed Pell's equation). As far as we know, such equations were first successfully treated by the Indian school. It is not known whether Archimedes himself had a method of solution.

Diophantus

Title page of the 1621 edition of Diophantus' *Arithmetica*, translated into Latin by Claude Gaspard Bachet de Méziriac.

Very little is known about Diophantus of Alexandria; he probably lived in the third century CE, that is, about five hundred years after Euclid. Six out of the thirteen books of Diophantus's *Arithmetica* survive in the original Greek; four more books survive in an Arabic translation. The *Arithmetica* is a collection of worked-out problems where the task is invariably to find rational solutions to a system of polynomial equations, usually of the form $f(x, y) = z^2$ or $f(x, y, z) = w^2$. Thus, nowadays, we speak of *Diophantine equations* when we speak of polynomial equations to which rational or integer solutions must be found.

One may say that Diophantus was studying rational points — i.e., points whose coordinates are rational — on curves and algebraic varieties; however, unlike the Greeks of the Classical period, who did what we would now call basic algebra in geometrical terms, Diophantus did what we would now call basic algebraic geometry in purely algebraic terms. In modern language, what Diophantus did was to find rational parametrizations of varieties; that is, given an equation of the form (say) $f(x_1, x_2, x_3) = 0$, his aim was to find (in essence) three rational functions g_1, g_2, g_3 such that, for all values of r and s, setting $x_i = g_i(r, s)$ for $i = 1, 2, 3$ gives a solution to $f(x_1, x_2, x_3) = 0$.

Diophantus also studied the equations of some non-rational curves, for which no rational parametrisation is possible. He managed to find some rational points on these curves (elliptic curves, as it happens, in what seems to be their first known occurrence) by means of what amounts to a tangent construction: translated into coordinate geometry (which did not exist in Diophantus's time), his method would be visualised as drawing a tangent to a curve at a known rational point, and then finding the other point of intersection of the tangent with the curve; that other point is a new rational point. (Diophantus also resorted to what could be called a special case of a secant construction.)

While Diophantus was concerned largely with rational solutions, he assumed some results on integer numbers, in particular that every integer is the sum of four squares (though he never stated as much explicitly).

Āryabhaṭa, Brahmagupta, Bhāskara

While Greek astronomy probably influenced Indian learning, to the point of introducing trigonometry, it seems to be the case that Indian mathematics is otherwise an indigenous tradition; in particular, there is no evidence that Euclid's Elements reached India before the 18th century.

Āryabhaṭa (476–550 CE) showed that pairs of simultaneous congruences $n \equiv a_1 \bmod m_1$, $n \equiv a_2 \bmod m_2$ could be solved by a method he called *kuṭṭaka*, or *pulveriser*; this is a procedure close to (a generalisation of) the Euclidean algorithm, which was probably discovered independently in India. Āryabhaṭa seems to have had in mind applications to astronomical calculations.

Brahmagupta (628 CE) started the systematic study of indefinite quadratic equations—in particular, the misnamed Pell equation, in which Archimedes may have first been interested, and which did not start to be solved in the West until the time of Fermat and Euler. Later Sanskrit authors would follow, using Brahmagupta's technical terminology. A general procedure (the chakravala, or "cyclic method") for solving Pell's equation was finally found by Jayadeva (cited in the eleventh century; his work is otherwise lost); the earliest surviving exposition appears in Bhāskara II's Bīja-gaṇita (twelfth century).

Indian mathematics remained largely unknown in Europe until the late eighteenth century; Brahmagupta and Bhāskara's work was translated into English in 1817 by Henry Colebrooke.

Arithmetic in the Islamic Golden Age

Al-Haytham seen by the West: frontispice of *Selenographia*, showing Alhasen [*sic*] representing knowledge through reason, and Galileo representing knowledge through the senses.

In the early ninth century, the caliph Al-Ma'mun ordered translations of many Greek mathematical works and at least one Sanskrit work (the *Sindhind*, which may or may not be Brahmagupta's Brāhmasphuṭasiddhānta). Diophantus's main work, the *Arithmetica*, was translated into Arabic by Qusta ibn Luqa (820–912). Part of the treatise *al-Fakhri* (by al-Karajī, 953 – ca. 1029) builds on it to some extent. According to Rashed Roshdi, Al-Karajī's contemporary Ibn al-Haytham knew what would later be called Wilson's theorem.

Western Europe in the Middle Ages

Other than a treatise on squares in arithmetic progression by Fibonacci — who lived and studied in north Africa and Constantinople during his formative years, ca. 1175–1200 — no number theory to speak of was done in western Europe during the Middle Ages. Matters started to change in Europe in the late Renaissance, thanks to a renewed study of the works of Greek antiquity. A catalyst was the textual emendation and translation into Latin of Diophantus's *Arithmetica* (Bachet, 1621, following a first attempt by Xylander, 1575).

Early Modern Number Theory

Fermat

Pierre de Fermat.

Pierre de Fermat (1601–1665) never published his writings; in particular, his work on number theory is contained almost entirely in letters to mathematicians and in private marginal notes. He wrote down nearly no proofs in number theory; he had no models in the area. He did make repeated use of mathematical induction, introducing the method of infinite descent.

One of Fermat's first interests was perfect numbers (which appear in Euclid, *Elements* IX) and amicable numbers; this led him to work on integer divisors, which were from the beginning among the subjects of the correspondence (1636 onwards) that put him in touch with the mathematical community of the day. He had already studied Bachet's edition of Diophantus carefully; by 1643, his interests had shifted largely to Diophantine problems and sums of squares (also treated by Diophantus).

Fermat's achievements in arithmetic include:

- Fermat's little theorem (1640), stating that, if a is not divisible by a prime p, then $a^{p-1} \equiv 1 \bmod p$.
- If a and b are coprime, then $a^2 + b^2$ is not divisible by any prime congruent to −1 modulo 4; and every prime congruent to 1 modulo 4 can be written in the form $a^2 + b^2$. These two statements also date from 1640; in 1659, Fermat stated to Huygens that he had proven the latter statement by the method of infinite descent. Fermat and Frenicle also did some work (some of it erroneous) on other quadratic forms.
- Fermat posed the problem of solving $x^2 - Ny^2 = 1$ as a challenge to English mathematicians (1657). The problem was solved in a few months by Wallis and Brouncker. Fermat considered their solution valid, but pointed out they had provided an algorithm without a proof (as had Jayadeva and Bhaskara, though Fermat would never know this.) He states that a proof can be found by descent.
- Fermat developed methods for (doing what in our terms amounts to) finding points on curves of genus 0 and 1. As in Diophantus, there are many special procedures and what amounts to a tangent construction, but no use of a secant construction.
- Fermat states and proves (by descent) in the appendix to *Observations on Diophantus* (Obs. XLV) that $x^4 + y^4 = z^4$ has no non-trivial solutions in the integers. Fermat also mentioned to his correspondents that $x^3 + y^3 = z^3$ has no non-trivial solutions, and that this could be proven by descent. The first known proof is due to Euler (1753; indeed by descent).

Fermat's claim ("Fermat's last theorem") to have shown there are no solutions to $x^n + y^n = z^n$ for all $n \geq 3$ (the only known proof of which is beyond his methods) appears only in his annotations on the margin of his copy of Diophantus; he never claimed this to others and thus would have had no need to retract it if he found any mistake in his supposed proof.

Euler

Leonhard Euler.

The interest of Leonhard Euler (1707–1783) in number theory was first spurred in 1729, when a friend of his, the amateur Goldbach, pointed him towards some of Fermat's work on the subject. This has been called the "rebirth" of modern number theory, after Fermat's relative lack of success in getting his contemporaries' attention for the subject. Euler's work on number theory includes the following:

- *Proofs for Fermat's statements.* This includes Fermat's little theorem (generalised by Euler to non-prime moduli); the fact that $p = x^2 + y^2$ if and only if $p \equiv 1 \bmod 4$; initial work towards a proof that every integer is the sum of four squares (the first complete proof is by Joseph-Louis Lagrange (1770), soon improved by Euler himself); the lack of non-zero integer solutions to $x^4 + y^4 = z^2$ (implying the case *n=4* of Fermat's last theorem, the case *n=3* of which Euler also proved by a related method).
- *Pell's equation*, first misnamed by Euler. He wrote on the link between continued fractions and Pell's equation.
- *First steps towards analytic number theory.* In his work of sums of four squares, partitions, pentagonal numbers, and the distribution of prime numbers, Euler pioneered the use of what can be seen as analysis (in particular, infinite series) in number theory. Since he lived before the development of complex analysis, most of his work is restricted to the formal manipulation of power series. He did, however, do some very notable (though not fully rigorous) early work on what would later be called the Riemann zeta function.
- *Quadratic forms.* Following Fermat's lead, Euler did further research on the question of which primes can be expressed in the form $x^2 + Ny^2$, some of it prefiguring quadratic reciprocity.
- *Diophantine equations.* Euler worked on some Diophantine equations of genus 0 and 1. In particular, he studied Diophantus's work; he tried to systematise it, but the time was not yet ripe for such an endeavour – algebraic geometry was still in its infancy. He did notice there was a connection between Diophantine problems and elliptic integrals, whose study he had himself initiated.

Lagrange, Legendre and Gauss

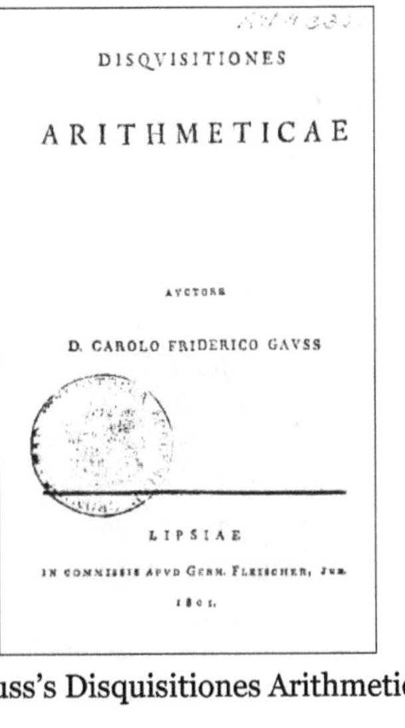
DISQVISITIONES

ARITHMETICAE

AVCTORE

D. CAROLO FRIDERICO GAVSS

LIPSIAE

IN COMMISSIS APVD GERH. FLEISCHER, IUN.

1801.

Carl Friedrich Gauss's Disquisitiones Arithmeticae, first edition.

Joseph-Louis Lagrange (1736–1813) was the first to give full proofs of some of Fermat's and Euler's work and observations – for instance, the four-square theorem and the basic theory of the misnamed "Pell's equation" (for which an algorithmic solution was found by Fermat and his contemporaries, and also by Jayadeva and Bhaskara II before them.) He also studied quadratic forms in full generality (as opposed to $mX^2 + nY^2$) — defining their equivalence relation, showing how to put them in reduced form, etc.

Adrien-Marie Legendre (1752–1833) was the first to state the law of quadratic reciprocity. He also conjectured what amounts to the prime number theorem and Dirichlet's theorem on arithmetic progressions. He gave a full treatment of the equation $ax^2 + by^2 + cz^2 = 0$ and worked on quadratic forms along the lines later developed fully by Gauss. In his old age, he was the first to prove "Fermat's last theorem" for $n = 5$ (completing work by Peter Gustav Lejeune Dirichlet, and crediting both him and Sophie Germain).

Carl Friedrich Gauss.

In his *Disquisitiones Arithmeticae* (1798), Carl Friedrich Gauss (1777–1855) proved the law of quadratic reciprocity and developed the theory of quadratic forms (in particular, defining their composition). He also introduced some basic notation (congruences) and devoted a section to computational matters, including primality tests. The last section of the *Disquisitiones* established a link between roots of unity and number theory:

The theory of the division of the circle...which is treated in sec. 7 does not belong by itself to arithmetic, but its principles can only be drawn from higher arithmetic.

In this way, Gauss arguably made a first foray towards both Évariste Galois's work and algebraic number theory.

Maturity and Division into Subfields

Ernst Kummer.

Peter Gustav Lejeune Dirichlet.

Starting early in the nineteenth century, the following developments gradually took place:

- The rise to self-consciousness of number theory (or *higher arithmetic*) as a field of study.
- The development of much of modern mathematics necessary for basic modern number theory: complex analysis, group theory, Galois theory—accompanied by greater rigor in analysis and abstraction in algebra.
- The rough subdivision of number theory into its modern subfields—in particular, analytic and algebraic number theory.

Algebraic number theory may be said to start with the study of reciprocity and cyclotomy, but truly came into its own with the development of abstract algebra and early ideal theory and valuation theory. A conventional starting point for analytic number theory is Dirichlet's theorem on arithmetic progressions (1837), whose proof introduced L-functions and involved some asymptotic analysis and a limiting process on a real variable. The first use of analytic ideas in number theory actually goes back to Euler (1730s), who used formal power series and non-rigorous (or implicit) limiting arguments. The use of *complex* analysis in number theory comes later: the work of Bernhard Riemann (1859) on the zeta function is the canonical starting point; Jacobi's four-square theorem (1839), which predates it, belongs to an initially different strand that has by now taken a leading role in analytic number theory (modular forms).

Main Subdivisions

Elementary Tools

The term *elementary* generally denotes a method that does not use complex analysis. For example, the prime number theorem was first proven using complex analysis in 1896, but an elementary proof was found only in 1949 by Erdős and Selberg. The term is somewhat ambiguous: for example, proofs based on complex Tauberian theorems (e.g. Wiener–Ikehara) are often seen as quite enlightening but not elementary, in spite of using Fourier analysis, rather than complex analysis as such. Here as elsewhere, an *elementary* proof may be longer and more difficult for most readers than a non-elementary one.

Number theory has the reputation of being a field many of whose results can be stated to the layperson. At the same time, the proofs of these results are not particularly accessible, in part because the range of tools they use is, if anything, unusually broad within mathematics.

Analytic Number Theory

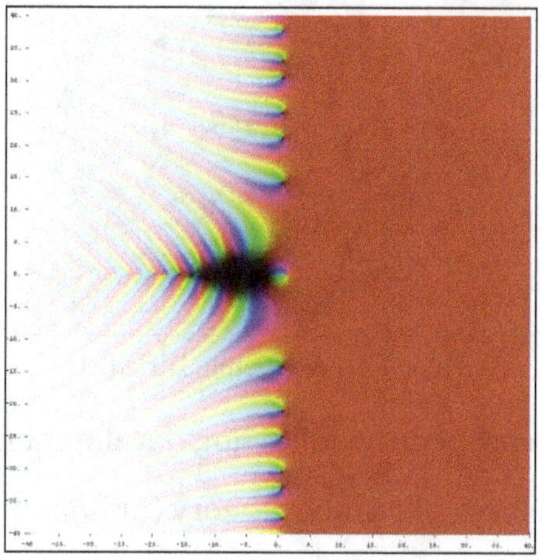

Riemann zeta function ζ(s) in the complex plane. The color of a point s gives the value of ζ(s): dark colors denote values close to zero and hue gives the value's argument.

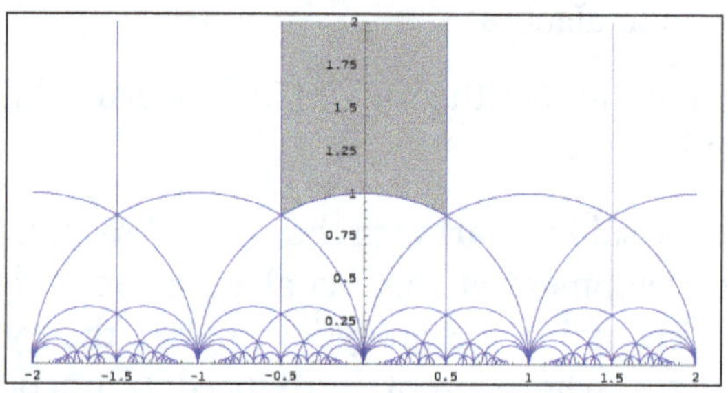

The action of the modular group on the upper half plane. The region in grey is the standard fundamental domain.

Analytic number theory may be defined,

- In terms of its tools, as the study of the integers by means of tools from real and complex analysis; or
- In terms of its concerns, as the study within number theory of estimates on size and density, as opposed to identities.

Some subjects generally considered to be part of analytic number theory, e.g., sieve theory, are better covered by the second rather than the first definition: some of sieve theory, for instance, uses little analysis, yet it does belong to analytic number theory.

The following are examples of problems in analytic number theory: the prime number theorem, the Goldbach conjecture (or the twin prime conjecture, or the Hardy–Littlewood conjectures), the Waring problem and the Riemann hypothesis. Some of the most important tools of analytic number theory are the circle method, sieve methods and L-functions (or, rather, the study of their properties). The theory of modular forms (and, more generally, automorphic forms) also occupies an increasingly central place in the toolbox of analytic number theory.

One may ask analytic questions about algebraic numbers, and use analytic means to answer such questions; it is thus that algebraic and analytic number theory intersect. For example, one may define prime ideals (generalizations of prime numbers in the field of algebraic numbers) and ask how many prime ideals there are up to a certain size. This question can be answered by means of an examination of Dedekind zeta functions, which are generalizations of the Riemann zeta function, a key analytic object at the roots of the subject. This is an example of a general procedure in analytic number theory: deriving information about the distribution of a sequence (here, prime ideals or prime numbers) from the analytic behavior of an appropriately constructed complex-valued function.

Algebraic Number Theory

An *algebraic number* is any complex number that is a solution to some polynomial equation $f(x)=0$ with rational coefficients; for example, every solution x of $x^5+(11/2)x^3-7x^2+9=0$ (say) is an algebraic number. Fields of algebraic numbers are also called *algebraic number ields*, or shortly *number ields*. Algebraic number theory studies algebraic number fields. Thus, analytic and algebraic number theory can and do overlap: the former is defined by its methods, the latter by its objects of study.

It could be argued that the simplest kind of number fields (viz., quadratic fields) were already studied by Gauss, as the discussion of quadratic forms in *Disquisitiones arithmeticae* can be restated in terms of ideals and norms in quadratic fields. (A *quadratic field* consists of all numbers of the form $a+b\sqrt{d}$, where a and b are rational numbers and d is a fixed rational number whose square root is not rational.) For that matter, the 11th-century chakravala method amounts—in modern terms—to an algorithm for finding the units of a real quadratic number field. However, neither Bhāskara nor Gauss knew of number fields as such.

The grounds of the subject as we know it were set in the late nineteenth century, when *ideal numbers*, the *theory of ideals* and *valuation theory* were developed; these are three complementary ways of dealing with the lack of unique factorisation in algebraic number fields. (For example, in the field generated by the rationals and $\sqrt{-5}$, the number 6 can be factorised both as $6=2\cdot 3$ and $6=(1+\sqrt{-5})(1-\sqrt{-5})$; all of 2, 2, $1+\sqrt{-5}$ and $1-\sqrt{-5}$ are irreducible, and thus, in a naïve sense, analogous to primes among the integers.) The initial impetus for the development of ideal numbers (by Kummer) seems to have come from the study of higher reciprocity laws,i.e., generalisations of quadratic reciprocity.

Number fields are often studied as extensions of smaller number fields: a field L is said to be an *extension* of a field K if L contains K. (For example, the complex numbers C are an extension of the reals R, and the reals R are an extension of the rationals Q.) Classifying the possible extensions of a given number field is a difficult and partially open problem. Abelian extensions—that is, extensions L of K such that the Galois group Gal(L/K) of L over K is an abelian group—are relatively well understood. Their classification was the object of the programme of class field theory, which was initiated in the late 19th century (partly by Kronecker and Eisenstein) and carried out largely in 1900—1950.

An example of an active area of research in algebraic number theory is Iwasawa theory. The Langlands program, one of the main current large-scale research plans in mathematics, is sometimes described as an attempt to generalise class field theory to non-abelian extensions of number fields.

Diophantine Geometry

The central problem of *Diophantine geometry* is to determine when a Diophantine equation has solutions, and if it does, how many. The approach taken is to think of the solutions of an equation as a geometric object.

For example, an equation in two variables defines a curve in the plane. More generally, an equation, or system of equations, in two or more variables defines a curve, a surface or some other such object in *n*-dimensional space. In Diophantine geometry, one asks whether there are any *rational points* (points all of whose coordinates are rationals) or *integral points* (points all of whose coordinates are integers) on the curve or surface. If there are any such points, the next step is to ask how many there are and how they are distributed. A basic question in this direction is: are there finitely or infinitely many rational points on a given curve (or surface)? What about integer points?

An example here may be helpful. Consider the Pythagorean equation $x^2 + y^2 = 1$; we would like to study its rational solutions, i.e., its solutions (x, y) such that x and y are both rational. This is the same as asking for all integer solutions to $a^2 + b^2 = c^2$; any solution to the latter equation gives us a solution $x = a/c$, $y = b/c$ to the former. It is also the same as asking for all points with rational coordinates on the curve described by $x^2 + y^2 = 1$. (This curve happens to be a circle of radius 1 around the origin.)

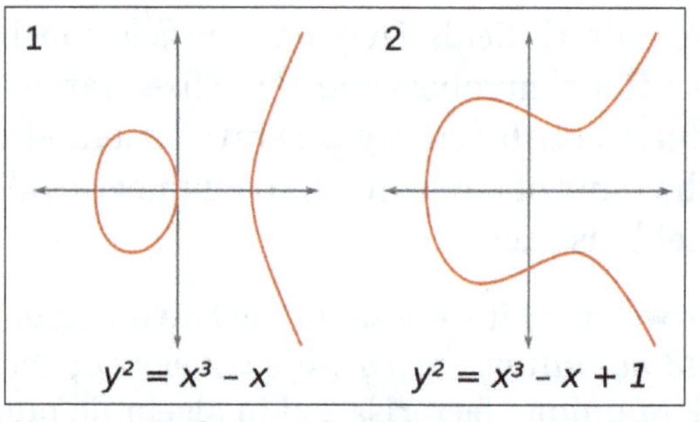

Two examples of an elliptic curve, i.e., a curve of genus 1 having at least one rational point. (Either graph can be seen as a slice of a torus in four-dimensional space.)

The rephrasing of questions on equations in terms of points on curves turns out to be felicitous. The finiteness or not of the number of rational or integer points on an algebraic curve—that is, rational or integer solutions to an equation $f(x, y) = 0$, where f is a polynomial in two variables—turns out to depend crucially on the *genus* of the curve. The *genus* can be defined as follows:[note 13] allow the variables in $f(x, y) = 0$ to be complex numbers; then $f(x, y) = 0$ defines a 2-dimensional surface in (projective) 4-dimensional space (since two complex variables can be decomposed into four real variables, i.e., four dimensions). Count the number of (doughnut) holes in the surface; call this number the *genus* of $f(x, y) = 0$. Other geometrical notions turn out to be just as crucial.

There is also the closely linked area of Diophantine approximations: given a number x, how well can it be approximated by rationals? (We are looking for approximations that are good relative to the amount of space that it takes to write the rational: call $a >$(with $\gcd(a, q) = 1$) a good approximation to x if $|x - a/q| < \frac{1}{q^c}$, where c is large.) This question is of special interest if x is an algebraic number. If x cannot be well approximated, then some equations do not have integer or rational solutions. Moreover, several concepts (especially that of height) turn out to be crucial both in Diophantine geometry and in the study of Diophantine approximations. This question is also of special interest in transcendental number theory: if a number can be better approximated than any algebraic number, then it is a transcendental number. It is by this argument that π and e have been shown to be transcendental.

Diophantine geometry should not be confused with the geometry of numbers, which is a collection of graphical methods for answering certain questions in algebraic number theory. *Arithmetic geometry*, on the other hand, is a contemporary term for much the same domain as that covered by the term *Diophantine geometry*. The term *arithmetic geometry* is arguably used most often when one wishes to emphasise the connections to modern algebraic geometry (as in, for instance, Faltings' theorem) rather than to techniques in Diophantine approximations.

Recent Approaches and Subfields

The areas below date as such from no earlier than the mid-twentieth century, even if they are based on older material. For example, as is explained below, the matter of algorithms in number theory is very old, in some sense older than the concept of proof; at the same time, the modern study of computability dates only from the 1930s and 1940s, and computational complexity theory from the 1970s.

Probabilistic Number Theory

Take a number at random between one and a million. How likely is it to be prime? This is just another way of asking how many primes there are between one and a million. Further: how many prime divisors will it have, on average? How many divisors will it have altogether, and with what likelihood? What is the probability that it will have many more or many fewer divisors or prime divisors than the average?

Much of probabilistic number theory can be seen as an important special case of the study of variables that are almost, but not quite, mutually independent. For example, the event that a random integer between one and a million be divisible by two and the event that it be divisible by three are almost independent, but not quite.

It is sometimes said that probabilistic combinatorics uses the fact that whatever happens with probability greater than 0 must happen sometimes; one may say with equal justice that many applications of probabilistic number theory hinge on the fact that whatever is unusual must be rare. If certain algebraic objects (say, rational or integer solutions to certain equations) can be shown to be in the tail of certain sensibly defined distributions, it follows that there must be few of them; this is a very concrete non-probabilistic statement following from a probabilistic one.

At times, a non-rigorous, probabilistic approach leads to a number of heuristic algorithms and open problems, notably Cramér's conjecture.

Arithmetic Combinatorics

Let A be a set of N integers. Consider the set $A + A = \{ m + n \mid m, n \in A \}$ consisting of all sums of two elements of A. Is $A + A$ much larger than A? Barely larger? If $A + A$ is barely larger than A, must A have plenty of arithmetic structure, for example, does A resemble an arithmetic progression?

If we begin from a fairly "thick" infinite set A, does it contain many elements in arithmetic progression: a, $a+b, a+2b, a+3b, \ldots, a+10b$, say? Should it be possible to write large integers as sums of elements of A?

These questions are characteristic of *arithmetic combinatorics*. This is a presently coalescing field; it subsumes *additive number theory* (which concerns itself with certain very specific sets A of arithmetic significance, such as the primes or the squares) and, arguably, some of the *geometry of numbers*, together with some rapidly developing new material. Its focus on issues of growth and distribution accounts in part for its developing links with ergodic theory, finite group theory, model theory, and other fields. The term *additive combinatorics* is also used; however, the sets A being studied need not be sets of integers, but rather subsets of non-commutative groups, for which the multiplication symbol, not the addition symbol, is traditionally used; they can also be subsets of rings, in which case the growth of $A + A$ and $A \cdot A$ may be compared.

Computations in Number Theory

While the word *algorithm* goes back only to certain readers of al-Khwārizmī, careful descriptions of methods of solution are older than proofs: such methods (that is, algorithms) are as old as any recognisable mathematics–ancient Egyptian, Babylonian, Vedic, Chinese–whereas proofs appeared only with the Greeks of the classical period. An interesting early case is that of what we now call the Euclidean algorithm. In its basic form (namely, as an algorithm for computing the greatest common divisor) it appears as Proposition 2 of Book VII in *Elements*, together with a proof of correctness. However, in the form that is often used in number theory (namely, as an algorithm for finding integer solutions to an equation $ax + by = c$, or, what is the same, for finding the quantities whose existence is assured by the Chinese remainder theorem) it first appears in the works of Āryabhaṭa (5th–6th century CE) as an algorithm called *kuṭṭaka* ("pulveriser"), without a proof of correctness.

There are two main questions: "can we compute this?" and "can we compute it rapidly?". Anybody can test whether a number is prime or, if it is not, split it into prime factors; doing so rapidly is another matter. We now know fast algorithms for testing primality, but, in spite of much work (both theoretical and practical), no truly fast algorithm for factoring.

The difficulty of a computation can be useful: modern protocols for encrypting messages (e.g., RSA) depend on functions that are known to all, but whose inverses (a) are known only to a chosen few, and (b) would take one too long a time to figure out on one's own. For example, these functions can be such that their inverses can be computed only if certain large integers are factorized. While many difficult computational problems outside number theory are known, most working encryption protocols nowadays are based on the difficulty of a few number-theoretical problems.

On a different note — some things may not be computable at all; in fact, this can be proven in some instances. For instance, in 1970, it was proven, as a solution to Hilbert's 10th problem, that there is no Turing machine which can solve all Diophantine equations. In particular, this means that, given a computably enumerable set of axioms, there are Diophantine equations for which there is no proof, starting from the axioms, of whether the set of equations has or does not have integer solutions. (We would necessarily be speaking of Diophantine equations for which there are no integer solutions, since, given a Diophantine equation with at least one solution, the solution itself provides a proof of the fact that a solution exists. We cannot prove, of course, that a particular Diophantine equation is of this kind, since this would imply that it has no solutions.)

Applications

The number-theorist Leonard Dickson (1874–1954) said "Thank God that number theory is unsullied by any application". Such a view is no longer applicable to number theory. In 1974, Donald Knuth said "...virtually every theorem in elementary number theory arises in a natural, motivated way in connection with the problem of making computers do high-speed numerical calculations". Elementary number theory is taught in discrete mathematics courses for computer scientists; and; on the other hand, number theory also has applications to the continuous in numerical analysis As well as the well-known applications to cryptography, there are also applications to many other areas of mathematics.

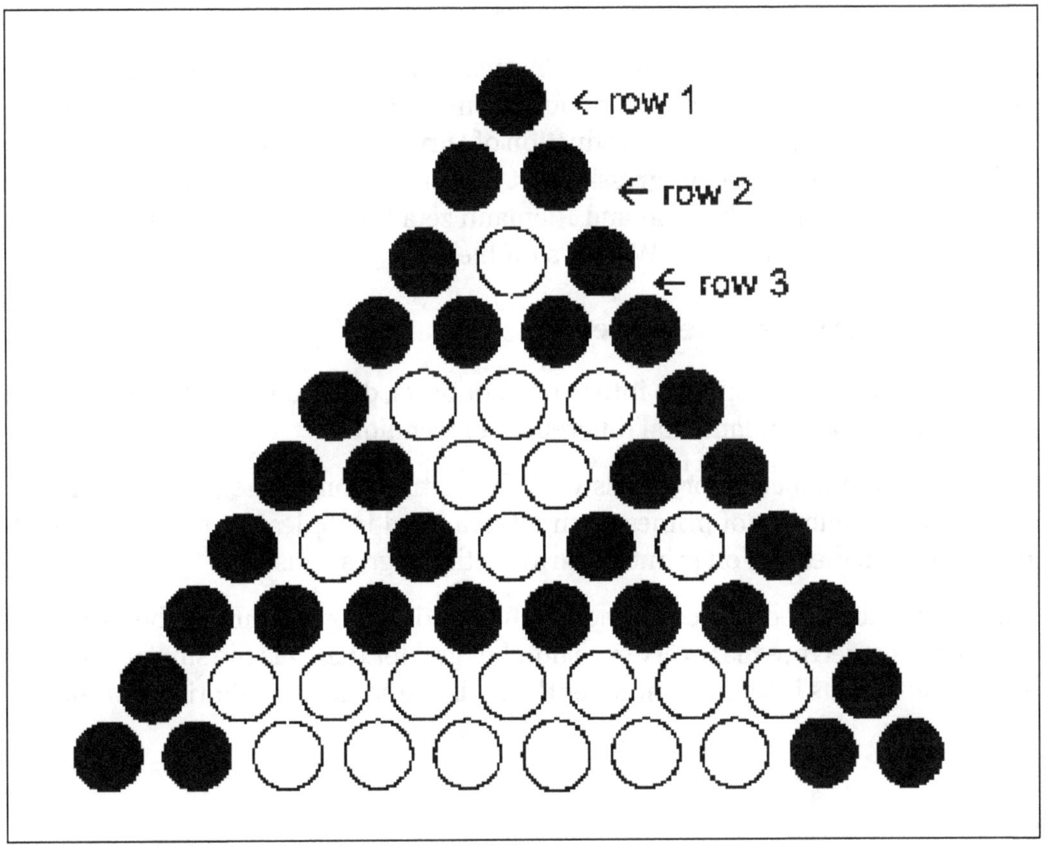

Prizes

The American Mathematical Society awards the *Cole Prize in Number Theory*. Moreover number theory is one of the three mathematical subdisciplines rewarded by the *Fermat Prize*.

Analytic Number Theory

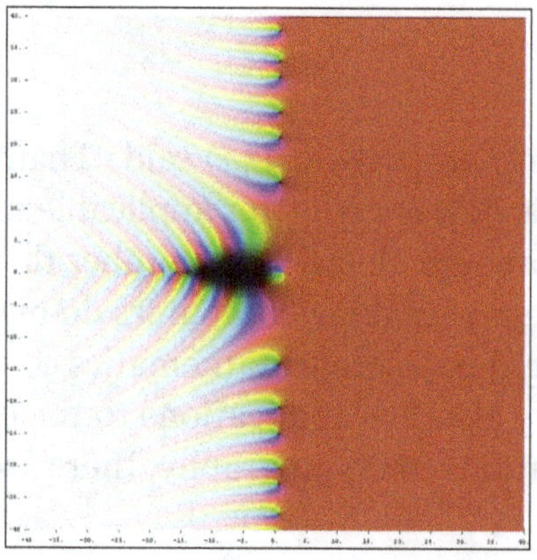

Riemann zeta function $\zeta(s)$ in the complex plane. The color of a point s encodes the value of $\zeta(s)$: colors close to black denote values close to zero, while hue encodes the value's argument.

In mathematics, analytic number theory is a branch of number theory that uses methods from mathematical analysis to solve problems about the integers. It is often said to have begun with Peter Gustav Lejeune Dirichlet's 1837 introduction of Dirichlet L-functions to give the first proof of Dirichlet's theorem on arithmetic progressions. It is well known for its results on prime numbers (involving the Prime Number Theorem and Riemann zeta function) and additive number theory (such as the Goldbach conjecture and Waring's problem).

Branches of Analytic Number Theory

Analytic number theory can be split up into two major parts, divided more by the type of problems they attempt to solve than fundamental differences in technique.

- Multiplicative number theory deals with the distribution of the prime numbers, such as estimating the number of primes in an interval, and includes the prime number theorem and Dirichlet's theorem on primes in arithmetic progressions.
- Additive number theory is concerned with the additive structure of the integers, such as Goldbach's conjecture that every even number greater than 2 is the sum of two primes. One of the main results in additive number theory is the solution to Waring's problem.

History

Precursors

Much of analytic number theory was inspired by the prime number theorem. Let $\pi(x)$ be the prime-counting function that gives the number of primes less than or equal to x, for any real number x.

For example, π(10) = 4 because there are four prime numbers (2, 3, 5 and 7) less than or equal to 10. The prime number theorem then states that $x / \ln(x)$ is a good approximation to π(x), in the sense that the limit of the *quotient* of the two functions π(x) and $x / \ln(x)$ as x approaches infinity is 1:

$$\lim_{x\to\infty} > \frac{\pi(x)}{x/\ln(x)} = 1$$

known as the asymptotic law of distribution of prime numbers.

Adrien-Marie Legendre conjectured in 1797 or 1798 that π(*a*) is approximated by the function $a/(A \ln(a) + B)$, where *A* and B are unspecified constants. In the second edition of his book on number theory (1808) he then made a more precise conjecture, with *A* = 1 and *B* = −1.08366. Carl Friedrich Gauss considered the same question: "Im Jahr 1792 oder 1793", according to his own recollection nearly sixty years later in a letter to Encke (1849), he wrote in his logarithm table (he was then 15 or 16) the short note "Primzahlen unter $a(=\infty)\frac{a}{\ln a}$". But Gauss never published this conjecture. In 1838 Peter Gustav Lejeune Dirichlet came up with his own approximating function, the logarithmic integral li(*x*) (under the slightly different form of a series, which he communicated to Gauss). Both Legendre's and Dirichlet's formulas imply the same conjectured asymptotic equivalence of π(*x*) and $x / \ln(x)$ stated above, although it turned out that Dirichlet's approximation is considerably better if one considers the differences instead of quotients.

Dirichlet

Johann Peter Gustav Lejeune Dirichlet is credited with the creation of analytic number theory, a field in which he found several deep results and in proving them introduced some fundamental tools, many of which were later named after him. In 1837 he published Dirichlet's theorem on arithmetic progressions, using mathematical analysis concepts to tackle an algebraic problem and thus creating the branch of analytic number theory. In proving the theorem, he introduced the Dirichlet characters and L-functions. In 1841 he generalized his arithmetic progressions theorem from integers to the ring of Gaussian integers $\mathbb{Z}[i]$.

Chebyshev

In two papers from 1848 and 1850, the Russian mathematician Pafnuty L'vovich Chebyshev attempted to prove the asymptotic law of distribution of prime numbers. His work is notable for the use of the zeta function ζ(*s*) (for real values of the argument "s", as are works of Leonhard Euler, as early as 1737) predating Riemann's celebrated memoir of 1859, and he succeeded in proving a slightly weaker form of the asymptotic law, namely, that if the limit of π(*x*)/(*x*/ln(*x*)) as *x* goes to infinity exists at all, then it is necessarily equal to one. He was able to prove unconditionally that this ratio is bounded above and below by two explicitly given constants near to 1 for all *x*. Although Chebyshev's paper did not prove the Prime Number Theorem, his estimates for π(*x*) were strong enough for him to prove Bertrand's postulate that there exists a prime number between *n* and 2*n* for any integer $n \geq 2$.

Riemann

"...es ist sehr wahrscheinlich, dass alle Wurzeln reell sind. Hiervon wäre allerdings ein strenger

Beweis zu wünschen; ich habe indess die Aufsuchung desselben nach einigen flüchtigen vergeblichen Versuchen vorläufig bei Seite gelassen, da er für den nächsten Zweck meiner Untersuchung entbehrlich schien."“...it is very probable that all roots are real. Of course one would wish for a rigorous proof here; I have for the time being, after some fleeting vain attempts, provisionally put aside the search for this, as it appears dispensable for the next objective of my investigation."

Riemann's statement of the Riemann hypothesis, from his 1859 paper. (He was discussing a version of the zeta function, modified so that its roots are real rather than on the critical line.)

Bernhard Riemann made some famous contributions to modern analytic number theory. In a single short paper (the only one he published on the subject of number theory), he investigated the Riemann zeta function and established its importance for understanding the distribution of prime numbers. He made a series of conjectures about properties of the zeta function, one of which is the well-known Riemann hypothesis.

Hadamard and De La Vallée-Poussin

xtending the ideas of Riemann, two proofs of the prime number theorem were obtained independently by Jacques Hadamard and Charles Jean de la Vallée-Poussin and appeared in the same year (1896). Both proofs used methods from complex analysis, establishing as a main step of the proof that the Riemann zeta function $\zeta(s)$ is non-zero for all complex values of the variable s that have the form $s = 1 + it$ with $t > 0$.

Modern Times

The biggest technical change after 1950 has been the development of *sieve methods*, particularly in multiplicative problems. These are combinatorial in nature, and quite varied. The extremal branch of combinatorial theory has in return been greatly influenced by the value placed in analytic number theory on quantitative upper and lower bounds. Another recent development is *probabilistic number theory*, which uses methods from probability theory to estimate the distribution of number theoretic functions, such as how many prime divisors a number has.

Developments within analytic number theory are often refinements of earlier techniques, which reduce the error terms and widen their applicability. For example, the *circle method* of Hardy and Littlewood was conceived as applying to power series near the unit circle in the complex plane; it is now thought of in terms of finite exponential sums (that is, on the unit circle, but with the power series truncated). The needs of diophantine approximation are for auxiliary functions that are not generating functions—their coefficients are constructed by use of a pigeonhole principle—and involve several complex variables. The fields of diophantine approximation and transcendence theory have expanded, to the point that the techniques have been applied to the Mordell conjecture.

Problems and Results

Theorems and results within analytic number theory tend not to be exact structural results about the integers, for which algebraic and geometrical tools are more appropriate. Instead, they give approximate bounds and estimates for various number theoretical functions, as the following examples illustrate.

Multiplicative Number Theory

Euclid showed that there are an infinite number of primes but it is very difficult to find an efficient method for determining whether or not a number is prime, especially a large number. A related but easier problem is to determine the asymptotic distribution of the prime numbers; that is, a rough description of how many primes are smaller than a given number. Gauss, amongst others, after computing a large list of primes, conjectured that the number of primes less than or equal to a large number N is close to the value of the integral.

$$\int_2^N \frac{1}{\log t} dt.$$

In 1859 Bernhard Riemann used complex analysis and a special meromorphic function now known as the Riemann zeta function to derive an analytic expression for the number of primes less than or equal to a real number x. Remarkably, the main term in Riemann's formula was exactly the above integral, lending substantial weight to Gauss's conjecture. Riemann found that the error terms in this expression, and hence the manner in which the primes are distributed, are closely related to the complex zeros of the zeta function. Using Riemann's ideas and by getting more information on the zeros of the zeta function, Jacques Hadamard and Charles Jean de la Vallée-Poussin managed to complete the proof of Gauss's conjecture. In particular, they proved that if,

$$\pi(x) = (\text{number of primes } \leq x),$$

then

$$\lim_{x \to \infty} \frac{\pi(x)}{x / \log x} = 1.$$

This remarkable result is what is now known as the *Prime Number Theorem*. It is a central result in analytic number theory. Loosely speaking, it states that given a large number N, the number of primes less than or equal to N is about $N/\log(N)$.

More generally, the same question can be asked about the number of primes in any arithmetic progression $a+nq$ for any integer n. In one of the first applications of analytic techniques to number theory, Dirichlet proved that any arithmetic progression with a and q coprime contains infinitely many primes. The prime number theorem can be generalised to this problem; letting

$$\pi(x,a,q) = (\text{number of primes } \leq x \text{ such that } p \text{ is in the arithmetic progression } a + nq, n \in \mathbf{Z}),$$

then if a and q are coprime,

$$\lim_{x \to \infty} \frac{\pi(x,a,q)\phi(q)}{x / \log x} = 1.$$

There are also many deep and wide ranging conjectures in number theory whose proofs seem too difficult for current techniques, such as the Twin prime conjecture which asks whether there are infinitely many primes p such that $p + 2$ is prime. On the assumption of the Elliott–Halberstam conjecture it has been proven recently that there are infinitely many primes p such that $p + k$ is prime for some positive even k at most 12. Also, it has been proven unconditionally (i.e. not depending on unproven conjectures) that there are infinitely many primes p such that $p + k$ is prime for some positive even k at most 246.

Additive Number Theory

One of the most important problems in additive number theory is Waring's problem, which asks whether it is possible, for any $k \geq 2$, to write any positive integer as the sum of a bounded number of kth powers,

$$n = x_1^k + \cdots + x_\ell^k.$$

The case for squares, $k = 2$, was answered by Lagrange in 1770, who proved that every positive integer is the sum of at most four squares. The general case was proved by Hilbert in 1909, using algebraic techniques which gave no explicit bounds. An important breakthrough was the application of analytic tools to the problem by Hardy and Littlewood. These techniques are known as the circle method, and give explicit upper bounds for the function $G(k)$, the smallest number of kth powers needed, such as Vinogradov's bound,

$$G(k) \leq k(3\log k + 11).$$

Diophantine Problems

Diophantine problems are concerned with integer solutions to polynomial equations: one may study the distribution of solutions, that is, counting solutions according to some measure of "size" or *height*.

An important example is the Gauss circle problem, which asks for integers points $(x\ y)$ which satisfy,

$$x^2 + y^2 \leq r^2.$$

In geometrical terms, given a circle centered about the origin in the plane with radius r, the problem asks how many integer lattice points lie on or inside the circle. It is not hard to prove that the answer is $\pi r^2 + E(r)$, where $E(r)/r^2 \to 0$ as $r \to \infty$. Again, the difficult part and a great achievement of analytic number theory is obtaining specific upper bounds on the error term $E(r)$.

It was shown by Gauss that $E(r) = O(r)$. In general, an $O(r)$ error term would be possible with the unit circle (or, more properly, the closed unit disk) replaced by the dilates of any bounded planar region with piecewise smooth boundary. Furthermore, replacing the unit circle by the unit square, the error term for the general problem can be as large as a linear function of r. Therefore, getting an error bound of the form $O(r^\delta)$ for some $\delta < 1$ in the case of the circle is a significant improvement. The first to attain this was Sierpiński in 1906, who showed $E(r) = O(r^{2/3})$. In 1915, Hardy and Landau each showed that one does *not* have $E(r) = O(r^{1/2})$. Since then the goal has been to show that for each fixed $\epsilon > 0$ there exists a real number $C(\epsilon)$ such that $E(r) \leq C(\epsilon) r^{1/2+\epsilon}$.

In 2000 Huxley showed that $E(r) = O(r^{131/208})$, which is the best published result.

Methods of Analytic Number Theory

Dirichlet Series

One of the most useful tools in multiplicative number theory are Dirichlet series, which are functions of a complex variable defined by an infinite series of the form,

$$f(s)=\sum_{n=1}^{\infty}a_n n^{-s}.$$

Depending on the choice of coefficients a_n, this series may converge everywhere, nowhere, or on some half plane. In many cases, even where the series does not converge everywhere, the holomorphic function it defines may be analytically continued to a meromorphic function on the entire complex plane. The utility of functions like this in multiplicative problems can be seen in the formal identity,

$$\left(\sum_{n=1}^{\infty}a_n n^{-s}\right)\left(\sum_{n=1}^{\infty}b_n n^{-s}\right)=\sum_{n=1}^{\infty}\left(\sum_{k\ell=n}a_k b_\ell\right)n^{-s};$$

hence the coefficients of the product of two Dirichlet series are the multiplicative convolutions of the original coefficients. Furthermore, techniques such as partial summation and Tauberian theorems can be used to get information about the coefficients from analytic information about the Dirichlet series. Thus a common method for estimating a multiplicative function is to express it as a Dirichlet series (or a product of simpler Dirichlet series using convolution identities), examine this series as a complex function and then convert this analytic information back into information about the original function.

Riemann Zeta Function

Euler showed that the fundamental theorem of arithmetic implies (at least formally) the *Euler product*,

$$\sum_{n=1}^{\infty}\frac{1}{n^s}=\prod_{p}^{\infty}\frac{1}{1-p^{-s}}\text{ for } s>1\ (p\text{ is prime number})$$

Euler's proof of the infinity of prime numbers makes use of the divergence of the term at the left hand side for $s = 1$ (the so-called harmonic series), a purely analytic result. Euler was also the first to use analytical arguments for the purpose of studying properties of integers, specifically by constructing generating power series. This was the beginning of analytic number theory.

Later, Riemann considered this function for complex values of s and showed that this function can be extended to a meromorphic function on the entire plane with a simple pole at $s = 1$. This function is now known as the Riemann Zeta function and is denoted by $\zeta(s)$. There is a plethora of literature on this function and the function is a special case of the more general Dirichlet L-functions.

Analytic number theorists are often interested in the error of approximations such as the prime number theorem. In this case, the error is smaller than $x/\log x$. Riemann's formula for $\pi(x)$ shows that the error term in this approximation can be expressed in terms of the zeros of the zeta function. In his 1859 paper, Riemann conjectured that all the "non-trivial" zeros of ζ lie on the line $\Re(s) = 1/2$ but never provided a proof of this statement. This famous and long-standing conjecture is known as the *Riemann Hypothesis* and has many deep implications in number theory; in fact, many important theorems have been proved under the assumption that the hypothesis is true. For example, under the assumption of the Riemann Hypothesis, the error term in the prime number theorem is $O(x^{1/2+\varepsilon})$.

In the early 20th century G. H. Hardy and Littlewood proved many results about the zeta function in an attempt to prove the Riemann Hypothesis. In fact, in 1914, Hardy proved that there were infinitely many zeros of the zeta function on the critical line,

$$\Re(z) = 1/2.$$

This led to several theorems describing the density of the zeros on the critical line.

Algebraic Number Theory

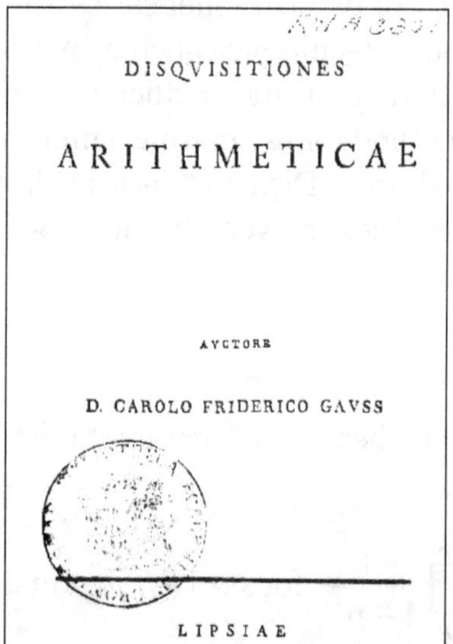

Title page of the first edition of Disquisitiones Arithmeticae, one of the founding works of modern algebraic number theory.

Algebraic number theory is a major branch of number theory that studies algebraic structures related to algebraic integers. This is generally accomplished by considering a ring of algebraic integers O in an algebraic number field K/Q, and studying their algebraic properties such as factorization, the behaviour of ideals, and field extensions. In this setting, the familiar features of the integers—such as unique factorization—need not hold. The virtue of the primary machinery employed—Galois theory, group cohomology, group representations, and L-functions—is that it allows one to deal with new phenomena and yet partially recover the behaviour of the usual integers.

History of Algebraic Number Theory

Diophantus

The beginnings of algebraic number theory can be traced to Diophantine equations, named after the 3rd-century Alexandrian mathematician, Diophantus, who studied them and developed methods for the solution of some kinds of Diophantine equations. A typical Diophantine problem is to find two integers x and y such that their sum, and the sum of their squares, equal two given numbers A and B, respectively:

$$A = x + y$$

$$B = x^2 + y^2.$$

Diophantine equations have been studied for thousands of years. For example, the solutions to the quadratic Diophantine equation $x^2 + y^2 = z^2$ are given by the Pythagorean triples, originally solved by the Babylonians (c. 1800 BC). Solutions to linear Diophantine equations, such as $26x + 65y = 13$, may be found using the Euclidean algorithm (c. 5th century BC).

Diophantus's major work was the *Arithmetica*, of which only a portion has survived.

Fermat

Fermat's last theorem was first conjectured by Pierre de Fermat in 1637, famously in the margin of a copy of *Arithmetica* where he claimed he had a proof that was too large to fit in the margin. No successful proof was published until 1995 despite the efforts of countless mathematicians during the 358 intervening years. The unsolved problem stimulated the development of algebraic number theory in the 19th century and the proof of the modularity theorem in the 20th century.

Gauss

One of the founding works of algebraic number theory, the *Disquisitiones Arithmeticae* (Latin: *Arithmetical Investigations*) is a textbook of number theory written in Latin by Carl Friedrich Gauss in 1798 when Gauss was 21 and first published in 1801 when he was 24. In this book Gauss brings together results in number theory obtained by mathematicians such as Fermat, Euler, Lagrange and Legendre and adds important new results of his own. Before the *Disquisitiones* was published, number theory consisted of a collection of isolated theorems and conjectures. Gauss brought the work of his predecessors together with his own original work into a systematic framework, filled in gaps, corrected unsound proofs, and extended the subject in numerous ways.

The *Disquisitiones* was the starting point for the work of other nineteenth century European mathematicians including Ernst Kummer, Peter Gustav Lejeune Dirichlet and Richard Dedekind. Many of the annotations given by Gauss are in effect announcements of further research of his own, some of which remained unpublished. They must have appeared particularly cryptic to his contemporaries; we can now read them as containing the germs of the theories of L-functions and complex multiplication, in particular.

Dirichlet

In a couple of papers in 1838 and 1839 Peter Gustav Lejeune Dirichlet proved the first class number formula, for quadratic forms (later refined by his student Kronecker). The formula, which Jacobi called a result "touching the utmost of human acumen", opened the way for similar results regarding more general number fields. Based on his research of the structure of the unit group of quadratic fields, he proved the Dirichlet unit theorem, a fundamental result in algebraic number theory.

He first used the pigeonhole principle, a basic counting argument, in the proof of a theorem in diophantine approximation, later named after him Dirichlet's approximation theorem. He published important contributions to Fermat's last theorem, for which he proved the cases $n = 5$ and $n = 14$, and to the biquadratic reciprocity law. The Dirichlet divisor problem, for which he found the first results, is still an unsolved problem in number theory despite later contributions by other researchers.

Dedekind

Richard Dedekind's study of Lejeune Dirichlet's work was what led him to his later study of algebraic number fields and ideals. In 1863, he published Lejeune Dirichlet's lectures on number theory as *Vorlesungen über Zahlentheorie* ("Lectures on Number Theory") about which it has been written that:

"Although the book is assuredly based on Dirichlet's lectures, and although Dedekind himself referred to the book throughout his life as Dirichlet's, the book itself was entirely written by Dedekind, for the most part after Dirichlet's death." (Edwards 1983)

1879 and 1894 editions of the *Vorlesungen* included supplements introducing the notion of an ideal, fundamental to ring theory. (The word "Ring", introduced later by Hilbert, does not appear in Dedekind's work.) Dedekind defined an ideal as a subset of a set of numbers, composed of algebraic integers that satisfy polynomial equations with integer coefficients. The concept underwent further development in the hands of Hilbert and, especially, of Emmy Noether. Ideals generalize Ernst Eduard Kummer's ideal numbers, devised as part of Kummer's 1843 attempt to prove Fermat's Last Theorem.

Hilbert

David Hilbert unified the field of algebraic number theory with his 1897 treatise *Zahlbericht* (literally "report on numbers"). He also resolved a significant number-theory problem formulated by Waring in 1770. As with the finiteness theorem, he used an existence proof that shows there must be solutions for the problem rather than providing a mechanism to produce the answers. He then had little more to publish on the subject; but the emergence of Hilbert modular forms in the dissertation of a student means his name is further attached to a major area.

He made a series of conjectures on class field theory. The concepts were highly influential, and his own contribution lives on in the names of the Hilbert class field and of the Hilbert symbol of local class field theory. Results were mostly proved by 1930, after work by Teiji Takagi.

Artin

Emil Artin established the Artin reciprocity law in a series of papers (1924; 1927; 1930). This law is a general theorem in number theory that forms a central part of global class field theory. The term "reciprocity law" refers to a long line of more concrete number theoretic statements which it generalized, from the quadratic reciprocity law and the reciprocity laws of Eisenstein and Kummer to Hilbert's product formula for the norm symbol. Artin's result provided a partial solution to Hilbert's ninth problem.

Modern Theory

Around 1955, Japanese mathematicians Goro Shimura and Yutaka Taniyama observed a possible link between two apparently completely distinct, branches of mathematics, elliptic curves and modular forms. The resulting modularity theorem (at the time known as the Taniyama–Shimura conjecture) states that every elliptic curve is modular, meaning that it can be associated with a unique modular form.

It was initially dismissed as unlikely or highly speculative, and was taken more seriously when number theorist André Weil found evidence supporting it, but no proof; as a result the "astounding" conjecture was often known as the Taniyama–Shimura-Weil conjecture. It became a part of the Langlands programme, a list of important conjectures needing proof or disproof.

From 1993 to 1994, Andrew Wiles provided a proof of the modularity theorem for semistable elliptic curves, which, together with Ribet's theorem, provides a proof for Fermat's Last Theorem. Both Fermat's Last Theorem and the Modularity Theorem were almost universally considered inaccessible to proof by contemporaneous mathematicians (meaning, impossible or virtually impossible to prove using current knowledge). Wiles first announced his proof in June 1993 in a version that was soon recognized as having a serious gap in a key point. The proof was corrected by Wiles, in part via collaboration with Richard Taylor, and the final, widely accepted, version was released in September 1994, and formally published in 1995. The proof uses many techniques from algebraic geometry and number theory, and has many ramifications in these branches of mathematics. It also uses standard constructions of modern algebraic geometry, such as the category of schemes and Iwasawa theory, and other 20th-century techniques not available to Fermat.

Basic Notions

Unique Factorization and the Ideal Class Group

One of the first properties of Z that can fail in the ring of integers *O* of an algebraic number field *K* is that of the unique factorization of integers into prime numbers. The prime numbers in Z are generalized to irreducible elements in *O*, and though the unique factorization of elements of *O* into irreducible elements may hold in some cases (such as for the Gaussian integers Z[i]), it may also fail, as in the case of Z[√-5] where

$$6 = 2 \cdot 3 = (1+\sqrt{-5}) \cdot (1-\sqrt{-5}).$$

The ideal class group of *O* is a measure of how much unique factorization of elements fails; in particular, the ideal class group is trivial if, and only if, *O* is a unique factorization domain.

Factoring Prime Ideals in Extensions

Unique factorization can be partially recovered for *O* in that it has the property of unique factorization of *ideals* into prime ideals (i.e. it is a Dedekind domain). This makes the study of the prime ideals in *O* particularly important. This is another area where things change from Z to *O*: the prime numbers, which generate prime ideals of Z (in fact, every single prime ideal of Z is of the form (*p*):=*p*Z for some prime number *p*,) may no longer generate prime ideals in *O*. For example, in the ring of Gaussian integers, the ideal 2Z[i] is no longer a prime ideal; in fact

$$2\mathbf{Z}[i] = ((1+i)\mathbf{Z}[i])^2$$

On the other hand, the ideal 3Z[i] is a prime ideal. The complete answer for the Gaussian integers is obtained by using a theorem of Fermat's, with the result being that for an odd prime number *p*,

$$p\mathbf{Z}[i] \text{ is a prime ideal if } p \equiv 3(\text{mod} 4)$$

$$p\mathbf{Z}[i] \text{ is not a prime ideal if } p \equiv 1(\text{mod} 4)$$

Generalizing this simple result to more general rings of integers is a basic problem in algebraic number theory. Class field theory accomplishes this goal when *K* is an abelian extension of Q (i.e. a Galois extension with abelian Galois group).

Primes and Places

An important generalization of the notion of prime ideal in *O* is obtained by passing from the so-called *ideal-theoretic* approach to the so-called *valuation-theoretic* approach. The relation between the two approaches arises as follows. In addition to the usual absolute value function $|\cdot| : Q \to R$, there are absolute value functions $|\cdot|_p : Q \to R$ defined for each prime number *p* in Z, called p-adic absolute values. Ostrowski's theorem states that these are all possible absolute value functions on Q (up to equivalence). This suggests that the usual absolute value could be considered as another prime. More generally, a prime of an algebraic number field ***K*** (also called a place) is an equivalence class of absolute values on *K*. The primes in *K* are of two sorts: $\mathfrak{p}$-adic absolute values like $|\cdot|_{\mathfrak{p}}$, one for each prime ideal $\mathfrak{p}$ of *O*, and absolute values like $|\cdot|$ obtained by considering *K* as a subset of the complex numbers in various possible ways and using the absolute value $|\cdot| : C \to R$. A prime of the first kind is called a finite prime (or finite place) and one of the second kind is called an infinite prime (or infinite place). Thus, the set of primes of Q is generally denoted { 2, 3, 5, 7, ..., ∞ }, and the usual absolute value on Q is often denoted $|\cdot|_\infty$ in this context.

The set of infinite primes of *K* can be described explicitly in terms of the embeddings $K \to C$ (i.e. the non-zero ring homomorphisms from *K* to C). Specifically, the set of embeddings can be split up into two disjoint subsets, those whose image is contained in R, and the rest. To each embedding $\sigma : K \to R$, there corresponds a unique prime of *K* coming from the absolute value obtained by composing σ with the usual absolute value on R; a prime arising in this fashion is called a real prime (or real place). To an embedding $\tau : K \to C$ whose image is *not* contained in R, one can construct a distinct embedding τ, called the *conjugate embedding*, by composing τ with the complex conjugation map $C \to C$. Given such a pair of embeddings τ and τ, there corresponds a unique prime of *K* again obtained by composing τ with the usual absolute value (composing τ instead gives the same absolute value function since |z| = |z| for any complex number z, where z denotes the complex conjugate of z). Such a prime is called a complex prime (or complex place). The description of the set of infinite primes is then as follows: each infinite prime corresponds either to a unique embedding $\sigma : K \to R$, or a pair of conjugate embeddings $\tau, \tau : K \to C$. The number of real (respectively, complex) primes is often denoted r_1 (respectively, r_2). Then, the total number of embeddings $K \to C$ is r_1+2r_2 (which, in fact, equals the degree of the extension K/Q).

Units

The fundamental theorem of arithmetic describes the multiplicative structure of Z. It states that every non-zero integer can be written (essentially) uniquely as a product of prime powers and ±1. The unique factorization of ideals in the ring *O* recovers part of this description, but fails to address the factor ±1. The integers 1 and -1 are the invertible elements (i.e. units) of Z. More generally, the invertible elements in *O* form a group under multiplication called the unit group of *O*, denoted $O^\times$. This group can be much larger than the cyclic group of order 2 formed by the units of Z. Dirichlet's unit theorem describes the abstract structure of the unit group as an abelian group. A more precise statement giving the structure of $O^\times \square_Z Q$ as a Galois module for the Galois group of K/Q is also possible. The size of the unit group, and its lattice structure give important numerical information about *O*, as can be seen in the class number formula.

Local Fields

Completing a number field *K* at a place *w* gives a complete field. If the valuation is archimedean, one gets R or C, if it is non-archimedean and lies over a prime *p* of the rationals, one gets a finite extension K_w / Q_p: a complete, discrete valued field with finite residue field. This process simplifies the arithmetic of the field and allows the local study of problems. For example, the Kronecker–Weber theorem can be deduced easily from the analogous local statement. The philosophy behind the study of local fields is largely motivated by geometric methods. In algebraic geometry, it is common to study varieties locally at a point by localizing to a maximal ideal. Global information can then be recovered by gluing together local data. This spirit is adopted in algebraic number theory. Given a prime in the ring of algebraic integers in a number field, it is desirable to study the field locally at that prime. Therefore, one localizes the ring of algebraic integers to that prime and then completes the fraction field much in the spirit of geometry.

Major Results

Finiteness of the Class Group

One of the classical results in algebraic number theory is that the ideal class group of an algebraic number field *K* is finite. The order of the class group is called the class number, and is often denoted by the letter *h*.

Dirichlet's Unit Theorem

Dirichlet's unit theorem provides a description of the structure of the multiplicative group of units $O^\times$ of the ring of integers *O*. Specifically, it states that $O^\times$ is isomorphic to $G \times Z^r$, where *G* is the finite cyclic group consisting of all the roots of unity in *O*, and $r = r_1 + r_2 - 1$ (where r_1 (respectively,

r_2) denotes the number of real embeddings (respectively, pairs of conjugate non-real embeddings) of K). In other words, $O^\times$ is a finitely generated abelian group of rank $r_1 + r_2 - 1$ whose torsion consists of the roots of unity in O.

Reciprocity Laws

In terms of the Legendre symbol, the law of quadratic reciprocity for positive odd primes states,

$$\left(\frac{p}{q}\right)\left(\frac{q}{p}\right) = (-1)^{\frac{p-1}{2}\frac{q-1}{2}}.$$

A reciprocity law is a generalization of the law of quadratic reciprocity.

There are several different ways to express reciprocity laws. The early reciprocity laws found in the 19th century were usually expressed in terms of a power residue symbol (p/q) generalizing the quadratic reciprocity symbol, that describes when a prime number is an nth power residue modulo another prime, and gave a relation between (p/q) and (q/p). Hilbert reformulated the reciprocity laws as saying that a product over p of Hilbert symbols $(a,b/p)$, taking values in roots of unity, is equal to 1. Artin's reformulated reciprocity law states that the Artin symbol from ideals (or ideles) to elements of a Galois group is trivial on a certain subgroup. Several more recent generalizations express reciprocity laws using cohomology of groups or representations of adelic groups or algebraic K-groups, and their relationship with the original quadratic reciprocity law can be hard to see.

Class Number Formula

The class number formula relates many important invariants of a number field to a special value of its Dedekind zeta function.

Related Areas

Algebraic number theory interacts with many other mathematical disciplines. It uses tools from homological algebra. Via the analogy of function fields vs. number fields, it relies on techniques and ideas from algebraic geometry. Moreover, the study of higher-dimensional schemes over Z instead of number rings is referred to as arithmetic geometry. Algebraic number theory is also used in the study of arithmetic hyperbolic 3-manifolds.

Algebraic Structure

In mathematics, and more specifically in abstract algebra, an algebraic structure is a set (called carrier set or underlying set) with one or more finitary operations defined on it that satisfies a list of axioms.

Examples of algebraic structures include groups, rings, fields, and lattices. More complex structures can be defined by introducing multiple operations, different underlying sets, or by altering the defining axioms. Examples of more complex algebraic structures include vector spaces, modules, and algebras.

The properties of specific algebraic structures are studied in abstract algebra. The general theory of algebraic structures has been formalized in universal algebra. Category theory is used to study the relationships between two or more classes of algebraic structures, often of different kinds. For example, Galois theory studies the connection between certain fields and groups, algebraic structures of two different kinds.

Introduction

Addition and multiplication on numbers are the prototypical example of an operation that combines two elements of a set to produce a third. These operations obey several algebraic laws. For example, $a+ (b + c) = (a + b) + c$ and $a(bc) = (ab)c$, both examples of the *associative law*. Also $a + b = b + a$, and $ab = ba$, the *commutative law*. Many systems studied by mathematicians have operations that obey some, but not necessarily all, of the laws of ordinary arithmetic. For example, rotations of objects in three-dimensional space can be combined by performing the first rotation and then applying the second rotation to the object in its new orientation. This operation on rotations obeys the associative law, but can fail the commutative law.

Mathematicians give names to sets with one or more operations that obey a particular collection of laws, and study them in the abstract as algebraic structures. When a new problem can be shown to follow the laws of one of these algebraic structures, all the work that has been done on that category in the past can be applied to the new problem.

In full generality, algebraic structures may involve an arbitrary number of sets and operations that can combine more than two elements (higher arity), but this article focuses on binary operations on one or two sets. The examples here are by no means a complete list, but they are meant to be a representative list and include the most common structures. Longer lists of algebraic structures may be found in the external links and within *Category:Algebraic structures*. Structures are listed in approximate order of increasing complexity.

Examples

One Set with Operations

Simple structures: no binary operation:

- Set: a degenerate algebraic structure S having no operations.
- Pointed set: S has one or more distinguished elements, often 0, 1, or both.
- Unary system: S and a single unary operation over S.
- Pointed unary system: a unary system with S a pointed set.

Group-like structures: one binary operation. The binary operation can be indicated by any symbol, or with no symbol (juxtaposition) as is done for ordinary multiplication of real numbers.

- Magma or groupoid: S and a single binary operation over S.
- Semigroup: an associative magma.
- Monoid: a semigroup with identity.

- Group: a monoid with a unary operation (inverse), giving rise to inverse elements.
- Abelian group: a group whose binary operation is commutative.
- Semilattice: a semigroup whose operation is idempotent and commutative. The binary operation can be called either meet or join.
- Quasigroup: a magma obeying the latin square property. A quasigroup may also be represented using three binary operations.
- Loop: a quasigroup with identity.

Ring-like structures or Ringoids: two binary operations, often called addition and multiplication, with multiplication distributing over addition.

- Semiring: a ringoid such that S is a monoid under each operation. Addition is typically assumed to be commutative and associative, and the monoid product is assumed to distribute over the addition on both sides, and the additive identity satisfies $0\,x = 0$ for all x.
- Near-ring: a semiring whose additive monoid is a (not necessarily abelian) group.
- Ring: a semiring whose additive monoid is an abelian group.
- Lie ring: a ringoid whose additive monoid is an abelian group, but whose multiplicative operation satisfies the Jacobi identity rather than associativity.
- Boolean ring: a commutative ring with idempotent multiplication operation.
- Field: a commutative ring which contains a multiplicative inverse for every nonzero element
- Kleene algebras: a semiring with idempotent addition and a unary operation, the Kleene star, satisfying additional properties.
- *-algebra: a ring with an additional unary operation (*) satisfying additional properties.

Lattice structures: two or more binary operations, including operations called meet and join, connected by the absorption law.

- Complete lattice: a lattice in which arbitrary meet and joins exist.
- Bounded lattice: a lattice with a greatest element and least element.
- Complemented lattice: a bounded lattice with a unary operation, complementation, denoted by postfix □. The join of an element with its complement is the greatest element, and the meet of the two elements is the least element.
- Modular lattice: a lattice whose elements satisfy the additional *modular identity*.
- Distributive lattice: a lattice in which each of meet and join distributes over the other. Distributive lattices are modular, but the converse does not hold.
- Boolean algebra: a complemented distributive lattice. Either of meet or join can be defined in terms of the other and complementation. This can be shown to be equivalent with the ring-like structure of the same name above.

- Heyting algebra: a bounded distributive lattice with an added binary operation, relative pseudo-complement, denoted by infix →, and governed by the axioms $x \to x = 1$, $x\,(x \to y) = x\,y$, $y\,(x \to y) = y$, $x \to (y\,z) = (x \to y)\,(x \to z)$.

Arithmetics: two binary operations, addition and multiplication. S is an infinite set. Arithmetics are pointed unary systems, whose unary operation is injective successor, and with distinguished element 0.

- Robinson arithmetic. Addition and multiplication are recursively defined by means of successor. 0 is the identity element for addition, and annihilates multiplication. Robinson arithmetic is listed here even though it is a variety, because of its closeness to Peano arithmetic.
- Peano arithmetic. Robinson arithmetic with an axiom schema of induction. Most ring and field axioms bearing on the properties of addition and multiplication are theorems of Peano arithmetic or of proper extensions thereof.

Two Sets with Operations

Module-like structures: composite systems involving two sets and employing at least two binary operations.

- Group with operators: a group G with a set Ω and a binary operation $\Omega \times G \to G$ satisfying certain axioms.
- Module: an Abelian group M and a ring R acting as operators on M. The members of R are sometimes called scalars, and the binary operation of *scalar multiplication* is a function $R \times M \to M$, which satisfies several axioms. Counting the ring operations these systems have at least three operations.
- Vector space: a module where the ring R is a division ring or field.
- Graded vector space: a vector space with a direct sum decomposition breaking the space into "grades".
- Quadratic space: a vector space V over a field F with a function from V into F satisfying certain properties. Every quadratic space is also an inner product space.

Algebra-like structures: composite system defined over two sets, a ring R and a R module M equipped with an operation called multiplication. This can be viewed as a system with five binary operations: two operations on R, two on M and one involving both R and M.

- Algebra over a ring (also *R-algebra*): a module over a commutative ring R, which also carries a multiplication operation that is compatible with the module structure. This includes distributivity over addition and linearity with respect to multiplication by elements of R. The theory of an algebra over a field is especially well developed.
- Associative algebra: an algebra over a ring such that the multiplication is associative.
- Nonassociative algebra: a module over a commutative ring, equipped with a ring multiplication operation that is not necessarily associative. Often associativity is replaced with a

different identity, such as alternation, the Jacobi identity, or the Jordan identity.

- Coalgebra: a vector space with a "comultiplication" defined dually to that of associative algebras.
- Lie algebra: a special type of nonassociative algebra whose product satisfies the Jacobi identity.
- Lie coalgebra: a vector space with a "comultiplication" defined dually to that of Lie algebras.
- Graded algebra: a graded vector space with an algebra structure compatible with the grading. The idea is that if the grades of two elements a and b are known, then the grade of ab is known, and so the location of the product ab is determined in the decomposition.
- Inner product space: an F vector space V with a bilinear binary operation from $V \times V \rightarrow F$.

Four or more binary operations:

- Bialgebra: an associative algebra with a compatible coalgebra structure.
- Lie bialgebra: a Lie algebra with a compatible bialgebra structure.
- Hopf algebra: a bialgebra with a connection axiom (antipode).
- Clifford algebra: a graded associative algebra equipped with an exterior product from which may be derived several possible inner products. Exterior algebras and geometric algebras are special cases of this construction.

Hybrid Structures

Algebraic structures can also coexist with added structure of non-algebraic nature, such as partial order or a topology. The added structure must be compatible, in some sense, with the algebraic structure.

- Topological group: a group with a topology compatible with the group operation.
- Lie group: a topological group with a compatible smooth manifold structure.
- Ordered groups, ordered rings and ordered fields: each type of structure with a compatible partial order.
- Archimedean group: a linearly ordered group for which the Archimedean property holds.
- Topological vector space: a vector space whose M has a compatible topology.
- Normed vector space: a vector space with a compatible norm. If such a space is complete (as a metric space) then it is called a Banach space.
- Hilbert space: an inner product space over the real or complex numbers whose inner product gives rise to a Banach space structure.
- Vertex operator algebra
- Von Neumann algebra: a *-algebra of operators on a Hilbert space equipped with the weak operator topology.

Universal Algebra

Algebraic structures are defined through different configurations of axioms. Universal algebra abstractly studies such objects. One major dichotomy is between structures that are axiomatized entirely by *identities* and structures that are not. If all axioms defining a class of algebras are identities, then the class of objects is a variety.

Identities are equations formulated using only the operations the structure allows, and variables that are tacitly universally quantified over the relevant universe. Identities contain no connectives, existentially quantified variables, or relations of any kind other than the allowed operations. The study of varieties is an important part of universal algebra. An algebraic structure in a variety may be understood as the quotient algebra of term algebra (also called "absolutely free algebra") divided by the equivalence relations generated by a set of identities. So, a collection of functions with given signatures generate a free algebra, the term algebra T. Given a set of equational identities (the axioms), one may consider their symmetric, transitive closure E. The quotient algebra T/E is then the algebraic structure or variety. Thus, for example, groups have a signature containing two operators: the multiplication operator m, taking two arguments, and the inverse operator i, taking one argument, and the identity element e, a constant, which may be considered an operator that takes zero arguments. Given a (countable) set of variables x, y, z, etc. the term algebra is the collection of all possible terms involving m, i, e and the variables; so for example, $m(i(x), m(x,m(y,e)))$ would be an element of the term algebra. One of the axioms defining a group is the identity $m(x, i(x)) = e$; another is $m(x,e) = x$. The axioms can be represented as trees. These equations induce equivalence classes on the free algebra; the quotient algebra then has the algebraic structure of a group.

Several non-variety structures fail to be varieties, because either:

1. It is necessary that $0 \neq 1$, 0 being the additive identity element and 1 being a multiplicative identity element, but this is a nonidentity.
2. Structures such as fields have some axioms that hold only for nonzero members of S. For an algebraic structure to be a variety, its operations must be defined for *all* members of S; there can be no partial operations.

Structures whose axioms unavoidably include nonidentities are among the most important ones in mathematics, e.g., fields and division rings. Although structures with nonidentities retain an undoubted algebraic flavor, they suffer from defects varieties do not have. For example, the product of two fields is not a field.

Category Theory

Category theory is another tool for studying algebraic structures (see, for example, Mac Lane 1998). A category is a collection of *objects* with associated *morphisms*. Every algebraic structure has its own notion of homomorphism, namely any function compatible with the operation(s) defining the structure. In this way, every algebraic structure gives rise to a category. For example, the category of groups has all groups as objects and all group homomorphisms as morphisms. This concrete category may be seen as a category of sets with added category-theoretic structure. Likewise, the category of topological groups (whose morphisms are the continuous group homomorphisms) is a category of topological spaces with extra structure. A forgetful functor between categories of algebraic structures "forgets" a part of a structure.

There are various concepts in category theory that try to capture the algebraic character of a context, for instance

- Algebraic Category.
- Essentially algebraic category.
- Presentable category.
- Locally presentable category.
- Monadic functors and categories.
- Universal property.

Different Meanings of "Structure"

In a slight abuse of notation, the word "structure" can also refer to just the operations on a structure, instead of the underlying set itself. For example, the sentence, "We have defined a ring *structure* on the set A," means that we have defined ring *operations* on the set A. For another example, the group $(\mathbb{Z},+)$ can be seen as a set $\mathbb{Z}$ that is equipped with an *algebraic structure,* namely the *operation* $+$.

Abstract Algebra

The permutations of Rubik's Cube form a group, a fundamental concept within abstract algebra.

In algebra, which is a broad division of mathematics, abstract algebra (occasionally called modern algebra) is the study of algebraic structures. Algebraic structures include groups, rings, fields, modules, vector spaces, lattices, and algebras. The term *abstract algebra* was coined in the early 20th century to distinguish this area of study from the other parts of algebra.

Algebraic structures, with their associated homomorphisms, form mathematical categories. Category theory is a formalism that allows a unified way for expressing properties and constructions that are similar for various structures.

Universal algebra is a related subject that studies types of algebraic structures as single objects. For example, the structure of groups is a single object in universal algebra, which is called *variety of groups*.

History

As in other parts of mathematics, concrete problems and examples have played important roles in the development of abstract algebra. Through the end of the nineteenth century, many -- perhaps most -- of these problems were in some way related to the theory of algebraic equations. Major themes include:

- Solving of systems of linear equations, which led to linear algebra.
- Attempts to find formulae for solutions of general polynomial equations of higher degree that resulted in discovery of groups as abstract manifestations of symmetry.
- Arithmetical investigations of quadratic and higher degree forms and diophantine equations, that directly produced the notions of a ring and ideal.

Numerous textbooks in abstract algebra start with axiomatic definitions of various algebraic structures and then proceed to establish their properties. This creates a false impression that in algebra axioms had come first and then served as a motivation and as a basis of further study. The true order of historical development was almost exactly the opposite. For example, the hypercomplex numbers of the nineteenth century had kinematic and physical motivations but challenged comprehension. Most theories that are now recognized as parts of algebra started as collections of disparate facts from various branches of mathematics, acquired a common theme that served as a core around which various results were grouped, and finally became unified on a basis of a common set of concepts. An archetypical example of this progressive synthesis can be seen in the history of group theory.

Early Group Theory

There were several threads in the early development of group theory, in modern language loosely corresponding to *number theory*, *theory of equations*, and *geometry*.

Leonhard Euler considered algebraic operations on numbers modulo an integer, modular arithmetic, in his generalization of Fermat's little theorem. These investigations were taken much further by Carl Friedrich Gauss, who considered the structure of multiplicative groups of residues mod n and established many properties of cyclic and more general abelian groups that arise in this way. In his investigations of composition of binary quadratic forms, Gauss explicitly stated the associative law for the composition of forms, but like Euler before him, he seems to have been more interested in concrete results than in general theory. In 1870, Leopold Kronecker gave a definition of an abelian group in the context of ideal class groups of a number field, generalizing Gauss's work; but it appears he did not tie his definition with previous work on groups, particularly permutation groups. In 1882, considering the same question, Heinrich M. Weber realized the connection and gave a similar definition that involved the cancellation property but omitted the existence of the inverse element, which was sufficient in his context (finite groups).

Permutations were studied by Joseph-Louis Lagrange in his 1770 paper *Réflexions sur la résolution algébrique des équations (Thoughts on the algebraic solution of equations)* devoted to solutions of algebraic equations, in which he introduced Lagrange resolvents. Lagrange's goal was to understand why equations of third and fourth degree admit formulae for solutions, and he identified as key objects permutations of the roots. An important novel step taken by Lagrange in this paper was the abstract view of the roots, i.e. as symbols and not as numbers. However, he did not consider composition of permutations. Serendipitously, the first edition of Edward Waring's *Meditationes Algebraicae* (*Meditations on Algebra*) appeared in the same year, with an expanded version published in 1782. Waring proved the main theorem on symmetric functions, and specially considered the relation between the roots of a quartic equation and its resolvent cubic. *Mémoire sur la résolution des équations* (*Memoire on the Solving of Equations*) of Alexandre Vandermonde (1771) developed the theory of symmetric functions from a slightly different angle, but like Lagrange, with the goal of understanding solvability of algebraic equations,

> *Kronecker claimed in 1888 that the study of modern algebra began with this first paper of Vandermonde. Cauchy states quite clearly that Vandermonde had priority over Lagrange for this remarkable idea, which eventually led to the study of group theory.*

Paolo Ruffini was the first person to develop the theory of permutation groups, and like his predecessors, also in the context of solving algebraic equations. His goal was to establish the impossibility of an algebraic solution to a general algebraic equation of degree greater than four. En route to this goal he introduced the notion of the order of an element of a group, conjugacy, the cycle decomposition of elements of permutation groups and the notions of primitive and imprimitive and proved some important theorems relating these concepts,

> *if G is a subgroup of S_5 whose order is divisible by 5 then G contains an element of order 5.*

Note, however, that he got by without formalizing the concept of a group, or even of a permutation group. The next step was taken by Évariste Galois in 1832, although his work remained unpublished until 1846, when he considered for the first time what is now called the *closure property* of a group of permutations, which he expressed as,

> ... if in such a group one has the substitutions S and T then one has the substitution ST.

The theory of permutation groups received further far-reaching development in the hands of Augustin Cauchy and Camille Jordan, both through introduction of new concepts and, primarily, a great wealth of results about special classes of permutation groups and even some general theorems. Among other things, Jordan defined a notion of isomorphism, still in the context of permutation groups and, incidentally, it was he who put the term *group* in wide use.

The abstract notion of a group appeared for the first time in Arthur Cayley's papers in 1854. Cayley realized that a group need not be a permutation group (or even *finite*), and may instead consist of matrices, whose algebraic properties, such as multiplication and inverses, he systematically investigated in succeeding years. Much later Cayley would revisit the question whether abstract groups were more general than permutation groups, and establish that, in fact, any group is isomorphic to a group of permutations.

Modern Algebra

The end of the 19th and the beginning of the 20th century saw a tremendous shift in the methodology of mathematics. Abstract algebra emerged around the start of the 20th century, under the name *modern algebra*. Its study was part of the drive for more intellectual rigor in mathematics. Initially, the assumptions in classical algebra, on which the whole of mathematics (and major parts of the natural sciences) depend, took the form of axiomatic systems. No longer satisfied with establishing properties of concrete objects, mathematicians started to turn their attention to general theory. Formal definitions of certain algebraic structures began to emerge in the 19th century. For example, results about various groups of permutations came to be seen as instances of general theorems that concern a general notion of an *abstract group*. Questions of structure and classification of various mathematical objects came to forefront.

These processes were occurring throughout all of mathematics, but became especially pronounced in algebra. Formal definition through primitive operations and axioms were proposed for many basic algebraic structures, such as groups, rings, and fields. Hence such things as group theory and ring theory took their places in pure mathematics. The algebraic investigations of general fields by Ernst Steinitz and of commutative and then general rings by David Hilbert, Emil Artin and Emmy Noether, building up on the work of Ernst Kummer, Leopold Kronecker and Richard Dedekind, who had considered ideals in commutative rings, and of Georg Frobenius and Issai Schur, concerning representation theory of groups, came to define abstract algebra. These developments of the last quarter of the 19th century and the first quarter of 20th century were systematically exposed in Bartel van der Waerden's *Moderne algebra*, the two-volume monograph published in 1930–1931 that forever changed for the mathematical world the meaning of the word *algebra* from *the theory of equations* to the *theory of algebraic structures*.

Basic Concepts

By abstracting away various amounts of detail, mathematicians have defined various algebraic structures that are used in many areas of mathematics. For instance, almost all systems studied are sets, to which the theorems of set theory apply. Those sets that have a certain binary operation defined on them form magmas, to which the concepts concerning magmas, as well those concerning sets, apply. We can add additional constraints on the algebraic structure, such as associativity (to form semigroups); identity, and inverses (to form groups); and other more complex structures. With additional structure, more theorems could be proved, but the generality is reduced. The "hierarchy" of algebraic objects (in terms of generality) creates a hierarchy of the corresponding theories: for instance, the theorems of group theory may be used when studying rings (algebraic objects that have two binary operations with certain axioms) since a ring is a group over one of its operations. In general there is a balance between the amount of generality and the richness of the theory: more general structures have usually less nontrivial theorems and less applications.

Examples of algebraic structures with a single binary operation are:

- Magmas
- Quasigroups
- Monoids

- Semigroups
- Groups

Examples involving several operations include:

- Rings
- Fields
- Modules
- Vector spaces
- Algebras over fields
- Associative algebras
- Lie algebras
- Lattices
- Boolean algebras

Applications

Because of its generality, abstract algebra is used in many fields of mathematics and science. For instance, algebraic topology uses algebraic objects to study topologies. The Poincaré conjecture, proved in 2003, asserts that the fundamental group of a manifold, which encodes information about connectedness, can be used to determine whether a manifold is a sphere or not. Algebraic number theory studies various number rings that generalize the set of integers. Using tools of algebraic number theory, Andrew Wiles proved Fermat's Last Theorem.

In physics, groups are used to represent symmetry operations, and the usage of group theory could simplify differential equations. In gauge theory, the requirement of local symmetry can be used to deduce the equations describing a system. The groups that describe those symmetries are Lie groups, and the study of Lie groups and Lie algebras reveals much about the physical system; for instance, the number of force carriers in a theory is equal to dimension of the Lie algebra, and these bosons interact with the force they mediate if the Lie algebra is nonabelian.

Algebraic Integer

In Algebraic number theory, an algebraic integer is a complex number that is a root of some monic polynomial (a polynomial whose leading coefficient is 1) with coefficients in $\mathbb{Z}$ (the set of integers). The set of all algebraic integers, A, is closed under addition and multiplication and therefore is a commutative subring of the complex numbers. The ring A is the integral closure of regular integers $\mathbb{Z}$ in complex numbers.

The ring of integers of a number field *K*, denoted by O_K, is the intersection of *K* and A: it can also be characterised as the maximal order of the field *K*. Each algebraic integer belongs to the ring of integers of some number field. A number x is an algebraic integer if and only if the ring $\mathbb{Z}[x]$ is finitely generated as an abelian group, which is to say, as a $\mathbb{Z}$-module.

Definitions

The following are equivalent definitions of an algebraic integer. Let K be a number field (i.e., a finite extension of $\mathbb{Q}$, the set of rational numbers), in other words, $K = \mathbb{Q}(\theta)$ for some algebraic number $\theta \in \mathbb{C}$ by the primitive element theorem.

- $\alpha \in K$ is an algebraic integer if there exists a monic polynomial $f(x) \in \mathbb{Z}[x]$ such that $f(\alpha) = 0$.
- $\alpha \in K$ is an algebraic integer if the minimal monic polynomial of α over $\mathbb{Q}$ is in $\mathbb{Z}[x]$.
- $\alpha \in K$ is an algebraic integer if $\mathbb{Z}[\alpha]$ is a finitely generated $\mathbb{Z}$-module.
- $\alpha \in K$ is an algebraic integer if there exists a finitely generated $\mathbb{Z}$-submodule $M \subset \mathbb{C}$ such that $\alpha M \subseteq M$.

Algebraic integers are a special case of integral elements of a ring extension. In particular, an algebraic integer is an integral element of a finite extension $K/\mathbb{Q}$.

Examples

- The only algebraic integers which are found in the set of rational numbers are the integers. In other words, the intersection of $\mathbb{Q}$ and A is exactly $\mathbb{Z}$. The rational number a/b is not an algebraic integer unless b divides a. Note that the leading coefficient of the polynomial $bx - a$ is the integer b. As another special case, the square root $\sqrt{n}$ of a non-negative integer n is an algebraic integer, and so is irrational unless n is a perfect square.
- If d is a square free integer then the extension $K = \mathbb{Q}(\sqrt{d})$ is a quadratic field of rational numbers. The ring of algebraic integers O_K contains $\sqrt{d}$ since this is a root of the monic polynomial $x^2 - d$. Moreover, if $d \equiv 1 \pmod 4$ the element $(1 + \sqrt{d})/2$ is also an algebraic integer. It satisfies the polynomial $x^2 - x + (1 - d)/4$ where the constant term $(1 - d)/4$ is an integer. The full ring of integers is generated by $\sqrt{d}$ or $(1 + \sqrt{d})/2$ respectively.
- The ring of integers of the field $F = \mathbb{Q}[\alpha], \alpha = \sqrt[3]{m}$ has the following integral basis, writing $m = hk^2$ for two square-free coprime integers h and k:

$$\begin{cases} 1, \alpha, \dfrac{\alpha^2 \pm k^2\alpha + k^2}{3k} & m \equiv \pm 1 \mod 9 \\ 1, \alpha, \dfrac{\alpha^2}{k} & \text{else} \end{cases}$$

- If ζ_n is a primitive n-th root of unity, then the ring of integers of the cyclotomic field $\mathbb{Q}(\zeta_n)$ is precisely $\mathbb{Z}[\zeta_n]$.
- If α is an algebraic integer then $\beta = \sqrt[n]{\alpha}$ is another algebraic integer. A polynomial for β is obtained by substituting x^n in the polynomial for α.

Non-example

- If $P(x)$ is a primitive polynomial which has integer coefficients but is not monic, and P is irreducible over $\mathbb{Q}$, then none of the roots of P are algebraic integers. (Here *primitive* is used in the sense that the highest common factor of the set of coefficients of P is 1; this is weaker than requiring the coefficients to be pairwise relatively prime.)

- is a Bézout domain, as a consequence of the principal ideal theorem.

Baker's Theorem

In transcendental number theory, a mathematical discipline, Baker's theorem gives a lower bound for the absolute value of linear combinations of logarithms of algebraic numbers. The result, proved by Alan Baker (1966, 1967a, 1967b), subsumed many earlier results in transcendental number theory and solved a problem posed by Alexander Gelfond nearly fifteen years earlier. Baker used this to prove the transcendence of many numbers, to derive effective bounds for the solutions of some Diophantine equations, and to solve the class number problem of finding all imaginary quadratic fields with class number 1.

History

To simplify notation, we introduce the set L of logarithms of nonzero algebraic numbers, that is

$$L=\left\{\lambda \in \mathbf{C}: e^{\lambda} \in \overline{\mathbf{Q}}\right\}.$$

Using this notation, several results in transcendental number theory become much easier to state. For example the Hermite–Lindemann theorem becomes the statement that any nonzero element of L is transcendental.

In 1934, Alexander Gelfond and Theodor Schneider independently proved the Gelfond–Schneider theorem. This result is usually stated as: if a is algebraic and not equal to 0 or 1, and if b is algebraic and irrational, then a^b is transcendental. Equivalently, though, it says that if λ_1 and λ_2 are elements of L that are linearly independent over the rational numbers, then they are linearly independent over the algebraic numbers. So if λ_1 and λ_2 are elements of L and λ_2 isn't zero, then the quotient λ_1/λ_2 is either a rational number or transcendental. It can't be an algebraic irrational number like $\sqrt{2}$.

Although proving this result of "rational linear independence implies algebraic linear independence" for two elements of L was sufficient for his and Schneider's result, Gelfond felt that it was crucial to extend this result to arbitrarily many elements of L. Indeed, from Gel'fond (1960, p. 177):

...one may assume ... that the most pressing problem in the theory of transcendental numbers is the investigation of the measures of transcendence of finite sets of logarithms of algebraic numbers.

This problem was solved fourteen years later by Alan Baker and has since had numerous applications not only to transcendence theory but in algebraic number theory and the study of Diophantine equations as well. Baker received the Fields medal in 1970 for both this work and his applications of it to Diophantine equations.

Statement

With the above notation, Baker's theorem is a nonhomogeneous generalisation of the Gelfond–Schneider theorem. Specifically it states:

Baker's Theorem. If $\lambda_1,\ldots,\lambda_n$ are elements of L that are linearly independent over the rational numbers, then for any algebraic numbers $\beta_0, \ldots, \beta_n$, not all zero, we have,

$$|\beta_0 + \beta_1\lambda_1 + \cdots + \beta_n\lambda_n| > H^{-C}$$

where H is the maximum of the heights of the β's and C is an effectively computable number depending on n, the λ's, and the maximum d of the degrees of the β's. (If β_0 is nonzero then the assumption that the λ's are linearly independent can be dropped.) In particular this number is nonzero, so 1 and the λ's are linearly independent over the algebraic numbers.

Just as the Gelfond–Schneider theorem is equivalent to the statement about the transcendence of numbers of the form a^b, so too Baker's theorem implies the transcendence of numbers of the form,

$$a_1^{b_1} \cdots a_n^{b_n},$$

where the b_i are all algebraic, irrational, and 1, $b_1,\ldots,b_n$ are linearly independent over the rationals, and the a_i are all algebraic and not 0 or 1.

Baker (1977) also gave several versions with explicit constants. For example, if $e^{\lambda_j} = \alpha_j$ has height at most $A_j \geq 4$ and all the numbers β_j have height at most $B \geq 4$ then the linear form,

$$\Lambda = \beta_0 + \beta_1\lambda_1 + \cdots + \beta_n\lambda_n$$

is either 0 or satisfies,

$$\log|\Lambda| > (16nd)^{200n}\Omega(\log\Omega - \log\log A_n)(\log B + \log\Omega)$$

where

$$\Omega = \log A_1 \log A_2 \cdots \log A_n$$

and the field generated by all the α's and β's over the rationals has degree at most d. In the special case when $\beta_0=0$ and all the β_j are rational integers, the rightmost term log Ω can be deleted.

An explicit result by Baker and Wüstholz for a linear form Λ with integer coefficients yields a lower bound of the form,

$$\log|\Lambda| > -Ch(\alpha_1)h(\alpha_2)\cdots h(\alpha_n)\log\left(\max\{|\beta_1|,\ldots,|\beta_n|\}\right),$$

with a constant C,

$$C = 18(n+1)! \cdot n^{n+1} \cdot (32d)^{n+2} \log(2nd),$$

where d is the degree of the number field generated by the α_i.

Baker's Method

Baker's proof of his theorem is an extension of the argument given by Gel'fond (1960, chapter III, section 4). The main ideas of the proof are illustrated by the proof of the following qualitative version of the theorem of Baker (1966) described by Serre (1971): if the numbers $2\pi i$ and $\log a_1, \ldots, \log a_n$ are linearly independent over the rational numbers, for nonzero algebraic numbers $a_1, \ldots, a_n$, then they are linearly independent over the algebraic numbers. The precise quantitative version of Bakers theory can be proved by replacing the conditions that things are zero by conditions that things are sufficiently small throughout the proof.

The main idea of Bakers proof is to construct an auxiliary function $\Phi(z_1,\ldots,z_{n-1})$ of several variables that vanishes to high order at many points of the form $\Phi(l,l,\ldots,l)$, then repeatedly show that it vanishes to lower order at even more points of this form. Finally the fact that it vanishes (to order 1) at enough points of this form implies using Vandermonde determinants that there is a multiplicative relation between the numbers a_i.

Construction of the Function Φ

Assume there is a relation,

$$\beta_1 \log \alpha_1 + \cdots + \beta_{n-1} \log \alpha_{n-1} = \log \alpha_n$$

for algebraic numbers $\alpha_1, \ldots, \alpha_n, \beta_1, \ldots, \beta_{n-1}$ The function Φ is of the form

$$\Phi(z_1,\ldots,z_{n-1}) = \sum_{\lambda_1=0}^{L} \cdots \sum_{\lambda_n=0}^{L} p(\lambda_1 \ldots, \lambda_n) \alpha_1^{(\lambda_1+\lambda_n\beta_1)z_1} \cdots \alpha_{n-1}^{(\lambda_{n-1}+\lambda_n\beta_{n-1})z_{n-1}}$$

The integer coefficients p are chosen so that they are not all zero and Φ and its derivatives of order at most some constant M vanish at $z_1 = \ldots = z_{n-1} = l$, for integers l with $0 \le l \le h$ for some constant h. This is possible because these conditions are homogeneous linear equations in the coefficients p, which have a non-zero solution provided the number of unknown variables p is larger than the number of equations. The linear relation between the logs of the α's is needed to cut down the number of linear equations that have to be satisfied. Moreover, using Siegel's lemma, the sizes of the coefficients p can be chosen to be not too large. The constants L, h, and M have to be carefully adjusted to that the next part of the proof works, and are subject to some constraints, which are roughly:

- L must be somewhat smaller than M to make the argument about extra zeros below work.
- A small power of h must be larger than L to make the final step of the proof work.
- L^n must be larger than about $M^{n-1}h$ in order that it is possible to solve for the coefficients p.

The constraints can be satisfied by taking h to be sufficiently large, M to be some fixed power of h, and L to be a slightly smaller power of h. Baker took M to be about h^2 and L to be about $h^{2-1/2n}$.

The linear relation between the logarithms of the α's is used to reduce L slightly; roughly speaking, without it the condition L^n must be larger than about $M^{n-1}h$ would become L^n must be larger than about $M^n h$, which is incompatible with the condition that L is somewhat smaller than M.

Zeros of Φ(l, ..., l)

The next step is to show that Φ vanishes to slightly smaller order at many more points of the form $z_1 = ... = z_{n-1} = l$ for integers l. This idea was Baker's key innovation: previous work on this problem involved trying to increase the number of derivatives that vanish while keeping the number of points fixed, which does not seem to work in the multivariable case. This is done by combining two ideas; First one shows that the derivatives at these points are quite small, by using the fact that many derivatives of Φ vanish at many nearby points. Then one shows that derivatives of Φ at this point are given by algebraic integers times known constants. If an algebraic integer has all its conjugates bounded by a known constant, then it cannot be too small unless it is zero, because the product of all conjugates of a nonzero algebraic integer is at least 1 in absolute value. Combining these two ideas implies that Φ vanishes to slightly smaller order at many more points $z_1 = ... = z_{n-1} = l$. This part of the argument requires that Φ does not increase too rapidly; the growth of Φ depends on the size of L, so requires a bound on the size of L, which turns out to be roughly that L must be somewhat smaller than M. More precisely, Baker showed that since Φ vanishes to order M at h consecutive integers, it also vanishes to order $M/2$ at $h^{1+1/8n}$ consecutive integers 1, 2, 3, ... Repeating this argument J times shows that Φ vanishes to order $M/2^J$ at $h^{1+J/8n}$ points, provided that h is sufficiently large and L is somewhat smaller than $M/2^J$.

One then takes J large enough that $h^{1+J/8n} > (L+1)^n$ (J larger than about $16n$ will do if $h^2 > L$) so that that $\Phi(l, ..., l) = 0$ for all integers l with $1 \le l \le (L+1)^n$.

Completion of the Proof

The condition that $\Phi(l, ..., l)=0$ for all integers l with $1\le l\le (L+1)^n$ can be written as,

$$\sum_{\lambda_1=0}^{L}\cdots\sum_{\lambda_n=0}^{L} p(\lambda_1\ldots,\lambda_n)\alpha_1^{\lambda_1 l}\cdots\alpha_n^{\lambda_n l}=0$$

This consists of $(L+1)^n$ homogeneous linear equations in the $(L+1)^n$ unknowns p, and by assumption has a non-zero solution p, so the determinant of the matrix of coefficients must vanish. However this matrix is a Vandermonde matrix, so the formula for the determinant of such a matrix forces an equality two of the values,

$$\alpha_1^{\lambda_1}\cdots\alpha_n^{\lambda_n}$$

so the numbers $\alpha_1,...,\alpha_n$ are multiplicatively dependent. Taking logs then shows that $2\pi i$, log α_1,...,log α_n are linearly dependent over the rationals.

Extensions and Generalizations

Baker (1966) in fact gave a quantitative version of the theorem, giving effective lower bounds for the linear form in logarithms. This is done by a similar argument, except statements about something being zero are replaced by statements giving a small upper bound for it, and so on.

Baker (1967a) showed how to eliminate the assumption about $2\pi i$ in the theorem. This requires a modification of the final step of the proof. One shows that many derivatives of the function $\varphi(z) = \Phi(z, ..., z)$ vanish at $z=0$, by an argument similar to the one above. But these equations for the first $(L+1)^n$ derivatives again give a homogeneous set of linear equations for the coefficients p, so the determinant is zero, and is again a Vandermonde determinant, this time for the numbers $\lambda_1 \log \alpha_1 + ... + \lambda_n \log \alpha_n$. So two of these expressions must be the same which shows that $\log \alpha_1,...,\log \alpha_n$ are linearly dependent over the rationals.

Baker (1967b) gave an inhomogeneous version of the theorem, showing that $\beta_0+\beta_1\log(\alpha_1) + ... + \beta_n \log(\alpha_n)$ is nonzero for nonzero algebraic numbers $\beta_0, ..., \beta_n, \alpha_1, ..., \alpha_n$, and moreover giving an effective lower bound for it. The proof is similar to the homogeneous case: one can assume that,

$$\beta_0 + \beta_1 \log \alpha_1 + \cdots + \beta_{n-1} \log \alpha_{n-1} = \log \alpha_n$$

and one inserts an extra variable z_0 into Φ as follows:

$$\Phi(z_0,\ldots,z_{n-1}) = \sum_{\lambda_0=0}^{L} \cdots \sum_{\lambda_n=0}^{L} p(\lambda_0 \ldots,\lambda_n) z_0^{\lambda_0} e^{\lambda_n \beta_0 z_0} \alpha_1^{(\lambda_1+\lambda_n\beta_1)z_1} \cdots \alpha_{n-1}^{(\lambda_{n-1}+\lambda_n\beta_{n-1})z_{n-1}}$$

Corollaries

As mentioned above, the theorem includes numerous earlier transcendence results concerning the exponential function, such as the Hermite–Lindemann theorem and Gelfond–Schneider theorem. It is not quite as encompassing as the still unproven Schanuel's conjecture, and does not imply the six exponentials theorem nor, clearly, the still open four exponentials conjecture.

The main reason Gelfond desired an extension of his result was not just for a slew of new transcendental numbers. In 1935 he used the tools he had developed to prove the Gelfond–Schneider theorem to derive a lower bound for the quantity,

$$|\beta_1\lambda_1 + \beta_2\lambda_2|$$

where β_1 and β_2 are algebraic and λ_1 and λ_2 are in L. Baker's proof gave lower bounds for quantities like the above but with arbitrarily many terms, and he could use these bounds to develop effective means of tackling Diophantine equations and to solve Gauss' class number problem.

Extensions

Baker's theorem grants us the linear independence over the algebraic numbers of logarithms of algebraic numbers. This is weaker than proving their *algebraic* independence. So far no progress has been made on this problem at all. It has been conjectured that if $\lambda_1,...,\lambda_n$ are elements of L that are linearly independent over the rational numbers, then they are algebraically independent too.

This is a special case of Schanuel's conjecture, but so far it remains to be proved that there even exist two algebraic numbers whose logarithms are algebraically independent. Indeed, Baker's theorem rules out linear relations between logarithms of algebraic numbers unless there are trivial reasons for them; the next most simple case, that of ruling out homogeneous quadratic relations, is the still open four exponentials conjecture.

Similarly, extending the result to algebraic independence but in the p-adic setting, and using the p-adic logarithm function, remains an open problem. It is known that proving algebraic independence of linearly independent p-adic logarithms of algebraic p-adic numbers would prove Leopoldt's conjecture on the p-adic ranks of units of a number field.

Chinese Remainder Theorem

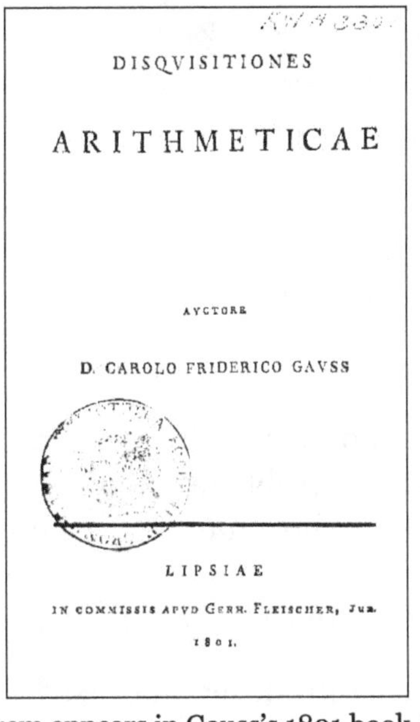
DISQVISITIONES

ARITHMETICAE

AVCTORE

D. CAROLO FRIDERICO GAVSS

LIPSIAE

IN COMMISSIS APVD GERH. FLEISCHER, Jun.

1801.

The Chinese remainder theorem appears in Gauss's 1801 book *Disquisitiones Arithmeticae.*

The Chinese remainder theorem is a theorem of number theory, which states that, if one knows the remainders of the division of an integer n by several integers, then one can determine uniquely the remainder of the division of n by the product of these integers, under the condition that the divisors are pairwise coprime.

This theorem has this name because it is a theorem about *remainders*, which was first discovered in the 3rd century AD by the Chinese mathematician Sunzi.

The Chinese remainder theorem is widely used for computing with large integers, as it allows replacing a computation for which one knows a bound on the size of the result by several similar computations on small integers.

The Chinese remainder theorem (expressed in terms of congruences) is true over every principal ideal domain. It has been generalized to any commutative ring, with a formulation involving ideals.

History

The earliest known statement of the theorem, as a problem with specific numbers, appears in the 3rd-century book *Sunzi's Mathematical Classic* by the Chinese mathematician Sun Tzu:

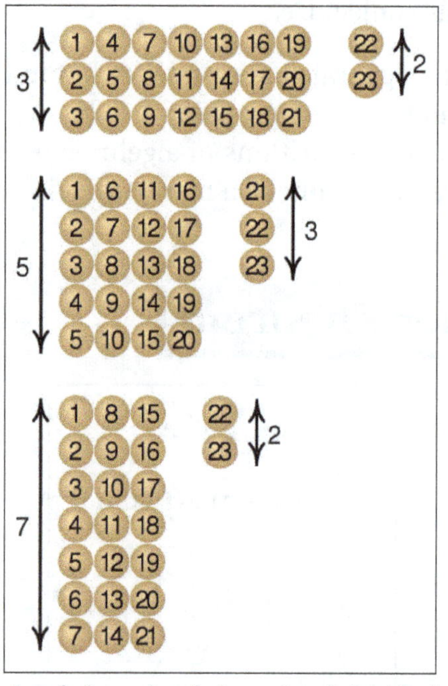

Sun Tzu's original formulation: $x \equiv 2 \pmod 3 \equiv 3 \pmod 5 \equiv 2 \pmod 7$ with the solution $x = 23 + 105k$ where $k \in \mathbb{Z}$.

Sun Tzu's work contains neither a proof nor a full algorithm. What amounts to an algorithm for solving this problem was described by Aryabhata (6th century). Special cases of the Chinese remainder theorem were also known to Brahmagupta (7th century), and appear in Fibonacci's Liber Abaci (1202). The result was later generalized with a complete solution called *Dayanshu* (大衍術) in Qin Jiushao's 1247 *Mathematical Treatise in Nine Sections* (數書九章, *Shushu Jiuzhang*).

The notion of congruences was first introduced and used by Gauss in his *Disquisitiones Arithmeticae* of 1801. Gauss illustrates the Chinese remainder theorem on a problem involving calendars, namely, "to find the years that have a certain period number with respect to the solar and lunar cycle and the Roman indiction." Gauss introduces a procedure for solving the problem that had already been used by Euler but was in fact an ancient method that had appeared several times.

Theorem Statement

Let $n_1, ..., n_k$ be integers greater than 1, which are often called *moduli* or *divisors*. Let us denote by N the product of the n_i.

The Chinese remainder theorem asserts that if the n_i are pairwise coprime, and if $a_1, ..., a_k$ are integers such that $0 \le a_i < n_i$ for every i, then there is one and only one integer x, such that $0 \le x < N$ and the remainder of the Euclidean division of x by n_i is a_i for every i.

This may be restated as follows in term of congruences: If the n_i are pairwise coprime, and if $a_1, ...,$

a_k are any integers, then there exists an integer x such that,

$$\begin{aligned} x &\equiv a_1 \pmod{n_1} \\ &\vdots \\ x &\equiv a_k \pmod{n_k}, \end{aligned}$$

and any two such x are congruent modulo N.

In abstract algebra, the theorem is often restated as: if the n_i are pairwise coprime, the map

$$x \bmod N \quad \mapsto \quad (x \bmod n_1, \ldots, x \bmod n_k)$$

defines a ring isomorphism,

$$\mathbb{Z}/N\mathbb{Z} \cong \mathbb{Z}/n_1\mathbb{Z} \times \cdots \times \mathbb{Z}/n_k\mathbb{Z}$$

between the ring of integers modulo N and the direct product of the rings of integers modulo the n_i. This means that for doing a sequence of arithmetic operations in $\mathbb{Z}/N\mathbb{Z}$, one may do the same computation independently in each $\mathbb{Z}/n_i\mathbb{Z}$ and then get the result by applying the isomorphism (from the right to the left). This may be much faster than the direct computation if N and the number of operations are large. This is widely used, under the name *multi-modular computation*, for linear algebra over the integers or the rational numbers.

The theorem can also be restated in the language of combinatorics as the fact that the infinite arithmetic progressions of integers form a Helly family.

Proof

The existence and the uniqueness of the solution may be proved independently. However the first proof of existence, given below, uses the uniqueness.

Uniqueness

Suppose that x and y are both solutions to all the congruences. As x and y give the same remainder, when divided by n_i, their difference $x - y$ is a multiple of each n_i. As the n_i are pairwise coprime, their product N divides also $x - y$, and thus x and y are congruent modulo N. If x and y are supposed to be non negative and less than N (as in the first statement of the theorem), then their difference may be a multiple of N only if $x = y$.

Existence (First Proof)

The map

$$x \mapsto (x \bmod n_1, \ldots, x \bmod n_k)$$

maps congruence classes modulo N to sequences of congruence classes modulo n_i. The proof of uniqueness shows that this map is injective. As the domain and the codomain of this map have the same number of elements, the map is also surjective, which proves the existence of the solution.

This proof is very simple, but does not provide any direct way for computing a solution. Moreover, it cannot be generalized to other situations where the following proof can.

Existence (Constructive Proof)

Existence may be established by an explicit construction of x. This construction may be split in two steps, firstly by solving the problem in the case of two moduli, and the second one by extending this solution to the general case by induction on the number of moduli.

Case of Two Moduli

We want to solve the system,

$$x \equiv a_1 \pmod{n_1}$$
$$x \equiv a_2 \pmod{n_2},$$

where n_1 and n_2 are coprime.

Bézout's identity asserts the existence of two integers m_1 and m_2 such that,

$$m_1 n_1 + m_2 n_2 = 1.$$

The integers m_1 and m_2 may be computed by Extended Euclidean algorithm.

A solution is given by,

$$x = a_1 m_2 n_2 + a_2 m_1 n_1.$$

In fact

$$\begin{aligned} x &= a_1 m_2 n_2 + a_2 m_1 n_1 \\ &= a_1 m_2 n_2 + a_2 m_1 n_1 + (a_1 m_1 n_1 - a_1 m_1 n_1) \\ &= a_1 (m_1 n_1 + m_2 n_2) - a_1 m_1 n_1 + a_2 m_1 n_1 \\ &= a_1 + n_1 m_1 (a_2 - a_1). \end{aligned}$$

This shows that $x \equiv a_1 \pmod{n_1}$. The second congruence is proved similarly, by exchanging the indices.

General Case

Let us consider a sequence of congruence equations,

$$x \equiv a_1 \pmod{n_1}$$
$$\vdots$$
$$x \equiv a_k \pmod{n_k},$$

where the n_i are pairwise coprime. The two first equations have a solution $a_{1,2}$ provided by the method of the previous section. The set of the solutions of these two first equations is the set of all solutions of the equation,

$$x \equiv a_{1,2} \pmod{n_1 n_2}.$$

As the other n_i are coprime with $n_1 n_2$, this reduces solving the initial problem of k equations to a similar problem with $k-1$ equations. Iterating the process, one gets eventually the solutions of the initial problem.

Existence (Direct Construction)

For constructing a solution, it is not necessary to make an induction on the number of moduli. However, such a direct construction involves more computation with large numbers, which makes it less efficient and less used. Nevertheless, Lagrange interpolation is a special case of this construction, applied to polynomials instead of integers.

Let $N_i = N / n_i$ be the product of all moduli but one. As the n_i are pairwise coprime, N_i and n_i are coprime. Thus Bézout's identity applies, and there exist integers M_i and m_i such that

$$M_i N_i + m_i n_i = 1.$$

A solution of the system of congruences is,

$$x = \sum_{i=1}^{k} a_i M_i N_i.$$

In fact, as N_i is a multiple of n_j for $i \neq j$, we have,

$$x \equiv a_i M_i N_i \equiv a_i (M_i N_i + m_i n_i) - a_i m_i n_i \equiv a_i \pmod{n_i},$$

for every i.

Computation

Let us consider a system of congruences,

$$x \equiv a_1 \pmod{n_1}$$
$$\vdots$$
$$x \equiv a_k \pmod{n_k},$$

where the n_i are pairwise coprime, and let $N = n_1, \ldots, n_k$. In this section several methods are described for computing the unique solution for x, such that $0 \leq x < N$, and these methods are applied on the example,

$$x \equiv 0 \pmod{3}$$
$$x \equiv 3 \pmod{4}$$
$$x \equiv 4 \pmod{5}.$$

Systematic Search

It is easy to check whether a value of x is a solution: it suffices to compute the remainder of the Euclidean division of x by each n_i. Thus, to find the solution, it suffices to check successively the integers from 0 to N until finding the solution.

Although very simple this method is very inefficient: for the simple example considered here, 40 integers (including 0 have to be checked for finding the solution 40. This is an exponential time algorithm, as the size of the input is, up to a constant factor, the number of digits of N, and the average number of operations is of the order of N.

Therefore, this method is rarely used, for hand-written computation as well on computers.

Search by Sieving

The search of the solution may be made dramatically faster by sieving. For this method, we suppose, without loss of generality, that $0 \le a_i < n_i$ (if it were not the case, it would suffice to replace each a_i by the remainder of its division by n_i). This implies that the solution belongs to the arithmetic progression,

$$a_1, a_1 + n_1, a_1 + 2n_1, \ldots$$

By testing the values of these numbers modulo n_2, one eventually finds a solution x_2 of the two first congruences. Then the solution belong to the arithmetic progression,

$$x_2, x_2 + n_1 n_2, x_2 + 2n_1 n_2, \ldots$$

Testing the values of these numbers modulo n_3,, and continuing until every modulus has been tested gives eventually the solution.

This method is faster if the moduli have been ordered by decreasing value, that is if $n_1 > n_2 > \cdots > n_k$. For the example, this gives the following computation. We consider first the numbers that are congruent to 4 modulo 5 (the largest modulus), which are 4, 9 = 4 + 5, 14 = 9 + 5, ... For each of them, compute the remainder by 4 (the second largest modulus) until getting a number congruent to 3 modulo 4. Then one can proceed by adding 20 = 5×4 at each step, and computing only the remainders by 3. This gives

4 mod 4 → 0. Continue

4 + 5 = 9 mod 4 →1. Continue

9 + 5 = 14 mod 4 → 2. Continue

14 + 5 = 19 mod 4 → 3. OK, continue by considering remainders modulo 3 and adding 5×4

= 20 each time

19 mod 3 → 1. Continue

19 + 20 = 39 mod 3 → 0. OK, this is the result.

This method works well for hand-written computation with a product of moduli that is not too big. However it is much slower than other methods, for very large products of moduli. Although dramatically faster than the systematic search, this method has also an exponential time complexity, and is therefore not used on computers.

Using the Existence Construction

The constructive existence proof shows that, in the case of two moduli, the solution may be obtained by the computation of the Bézout coefficients of the moduli, followed by a few multiplications, additions and reductions modulo n_1n_2 (for getting a result in the interval $(0, n_1n_2 - 1)$. As the Bézout's coefficients may be computed with the extended Euclidean algorithm, the whole computation, at most, has a quadratic time complexity of $O((s_1 + s_2)^2)$, where s_i denotes the number of digits of n_i.

For more than two moduli, the method for two moduli allows the replacement of any two congruences by a single congruence modulo the product of the moduli. Iterating this process provides eventually the solution with a complexity, which is quadratic in the number of digits of the product of all moduli. This quadratic time complexity does not depend on the order in which the moduli are regrouped. One may regroup the two first moduli, then regrouping the resulting modulus with the next one, and so on. This is the strategy which is the easiest to implement, but it needs more computation involving large numbers.

Another strategy consists in partitioning the moduli in pairs whose product have comparable sizes (as much as possible), applying parallely the method of two moduli to each pair, and iterating with a number of moduli approximatively divided by two. This method allows an easy parallelization of the algorithm. Also, if fast algorithms (that is algorithms working in quasilinear time) are used for the basic operations, this method provides an algorithm for the whole computation that works in quasilinear time.

On the current example (which has only three moduli), both strategies are identical, and works as follows.

Bézout's identity for 3 and 4 is,

$$1 \times 4 + (-1) \times 3 = 1.$$

Putting this in the formula given for proving the existence gives,

$$0 \times 4 + -3 \times 3 = -9$$

for a solution of the two first congruences, the other solutions being obtained by adding to −9 any multiple of 3×4 = 12. One may continue with any of these solutions, but the solution 3 = −9 +12 is smaller (in absolute value) and thus leads probably to an easier computation,

Bézout identity for 5 and 3×4 = 12 is,

$$5\times5+(-2)\times12=1.$$

Applying the same formula again, we get a solution of the problem:

$$25\times3-24\times4=-21.$$

The other solutions are obtained by adding any multiple of 3×4×5 = 60, and the smallest positive solution is −21 + 60 = 39.

As a Linear Diophantine System

The system of congruences solved by the Chinese remainder theorem may be rewritten as a system of simultaneous linear Diophantine equations,

$$\begin{aligned} x &= a_1 + x_1 n_1 \\ &\vdots \\ x &= a_k + x_k n_k, \end{aligned}$$

where the unknown integers are x and the x_i. Therefore, every general method for solving such systems may be used for finding the solution of Chinese remainder theorem, such as the reduction of the matrix of the system to Smith normal form or Hermite normal form. However, as usual when using a general algorithm for a more specific problem, this approach is less efficient than the method of the preceding section, based on a direct use of Bézout's identity.

Over Principal Ideal Domains

In § Theorem statement, the Chinese remainder theorem has been stated in three different ways: in terms of remainders, of congruences and of a ring isomorphism. The statement in terms of remainders does not apply, in general to principal ideal domains, as remainders are not defined in such rings. However, the two other version make sense over a principal ideal domain R: it suffices to replace "integer" by "element of the domain", and $\mathbb{Z}$ by R. These two versions of the theorem are true in this context, because the proofs (except for the first existence proof), are based on Euclid's lemma and Bézout's identity, which are true over every principal domain.

However, in general, the theorem is only an existence theorem, and does not provide any way for computing the solution, unless if one has an algorithm for computing the coefficients of Bézout's identity.

Over Univariate Polynomial Rings and Euclidean Domains

The statement in terms of remainders given in § Theorem statement cannot be generalized to any principal ideal domain, but its generalization to Euclidean domains is straightforward. The univariate polynomials over a field is the typical example of a Euclidean domain, which is not the integers. therefore, we state the theorem for the case of a ring of univariate domain $R = K[X]$ over a field K. For getting the theorem for a general Euclidean domain, it suffices to replace the degree by the Euclidean function of the Euclidean domain.

The Chinese remainder theorem for polynomials is thus: Let $P_i(X)$ (the moduli) be, for i=1, ..., k, pairwise coprime polynomials in $R = K[X]$. Let $d_i = \deg P_i$ be the degree of $P_i(X)$, and D be the sum of the d_i. If $A_1(X),\ldots,A_k(X)$ are polynomials such that $A_i(X)=0$ or $\deg A_i < d_i$ for every i, then, there is one and only one polynomial $P(X)$, such that $\deg P < D$ and the remainder of the Euclidean division of $P(X)$ by $P_i(X)$ is $A_i(X)$ for every i.

The construction of the solution may be done as in § Existence (constructive proof) or § Existence (direct proof). However, the latter construction may be simplified by using, as follows, partial fraction decomposition instead of extended Euclidean algorithm.

Thus, we want to find a polynomial $P(X)$, which satisfies the congruences,

$$P(X) \equiv A_i(X) \pmod{P_i(X)},$$

for $i = 1,\ldots,k$.

Let us consider the polynomials,

$$Q(X) = \prod_{i=1}^{k} P_i(X)$$

$$Q_i(X) = \frac{Q(X)}{P_i(X)}.$$

The partial fraction decomposition of $1/Q(X)$ gives k polynomials $S_i(X)$ with degrees $\deg S_i(X) < d_i$, such that

$$\frac{1}{Q(X)} = \sum_{i=1}^{k} \frac{S_i(X)}{P_i(X)},$$

and thus

$$1 = \sum_{i=1}^{k} S_i(X) Q_i(X).$$

Then a solution of the simultaneous congruence system is given by the polynomial,

$$\sum_{i=1}^{k} A_i(X) S_i(X) Q_i(X).$$

In fact, we have

$$\sum_{i=1}^{k} A_i(X) S_i(X) Q_i(X) = A_i(X) + \sum_{j=1}^{k} (A_j(X) - A_i(X)) S_j(X) Q_j(X) \equiv A_i(X) \pmod{P_i(X)},$$

for $1 \leq i \leq k$.

This solution may have a degree larger that vThe unique solution of degree less than D may be deduced by considering the remainder $B_i(X)$ of the Euclidean division of $A_i(X)S_i(X)$ by $P_i(X)$. This solution is,

$$P(X) = \sum_{i=1}^{k} B_i(X)Q_i(X).$$

Lagrange Interpolation

A special case of Chinese remainder theorem for polynomials is Lagrange interpolation. For this, let us consider k monic polynomials of degree one:

$$P_i(X) = X - x_i.$$

They are pairwise coprime if the x_i are all different. The remainder of the division by $P_i(X)$ of a polynomial $P(X)$ is $P(x_i)$.

Now, let $A_1, \ldots, A_k$ be constants (polynomials of degree 0) in K. Both Lagrange interpolation and Chinese remainder theorem assert the existence of a unique polynomial $P(X)$, of degree less than k such that,

$$P(x_i) = A_i,$$

for every i.

Lagrange interpolation formula is exactly the result, in this case, of the above construction of the solution. More precisely, let

$$Q(X) = \prod_{i=1}^{k} (X - x_i)$$

$$Q_i(X) = \frac{Q(X)}{X - x_i}.$$

The partial fraction decomposition of $1/Q(X)$ is,

$$\frac{1}{Q(X)} = \sum_{i=1}^{k} \frac{1}{Q_i(x_i)(X - X_i)}.$$

In fact, reducing the right-hand side to a common denominator one gets $\sum_{i=1}^{k} \frac{1}{Q_i(x_i)(X - X_i)} = \frac{\sum_{i=1}^{k} Q_i(X)/Q_i(x_i)}{Q(X)}$, and the numerator is equal to one, as being a polynomial of degree less than k, which takes the value one for k different values of X.

Using the above general formula, we get the Lagrange interpolation formula,

$$P(X)=\sum_{i=1}^{k}A_i\frac{Q_i(X)}{Q_i(x_i)}.$$

Hermite Interpolation

Hermite interpolation is an application of Chinese remainder theorem for univariate polynomials, which may involve moduli of arbitrary degrees (Lagrange interpolation involves only moduli of degree one).

The problem consists of finding a polynomial of the least possible degree, such that the polynomial and its first derivatives take given values at some fixed points.

More precisely, let $x_1,\ldots,x_k$ be k elements of the ground field K, and, for $i=1,\ldots,k$, let $a_{i,0},a_{i,1},\ldots,a_{i,r_i-1}$ be the values of the first r_i derivatives of the sought polynomial at x_i (including the 0th derivative, which is the value of the polynomial itself). The problem is to find a polynomial $P(X)$ such that its *j*th derivative takes the value $a_{i,j}$ at x_i, for $i=1,\ldots,k$ and $j=0,\ldots,r_j$.

Let us consider the polynomial,

$$P_i(X)=\sum_{j=0}^{r_i-1}\frac{a_{i,j}}{j!}(X-x_i)^j.$$

This is the Taylor polynomial of order r_i-1 at x_i,, of the unknown polynomial $P(X)$. Therefore, we must have,

$$P(X)\equiv P_i(X)\pmod{(X-x_i)^{r_i}}.$$

Conversely, any polynomial $P(X)$ that satisfies these k congruences, in particular verifies, for any $i=1,\ldots,k$

$$P(X)=P_i(X)+o(X-x_i)^{r_i-1}$$

therefore $P_i(X)$ is its Taylor polynomial of order r_i-1 at x_i, that is, $P(X)$ solves the initial Hermite interpolation problem. The Chinese remainder theorem asserts that there exists exactly one polynomial of degree less than the sum of the r_i, which satisfies these k congruences.

There are several ways for computing the solution $P(X)$. One may use the method described at the beginning of § Over univariate polynomial rings and Euclidean domains. One may also use the constructions given in § Existence (constructive proof) or § Existence (direct proof).

Generalization to Arbitrary Rings

The Chinese remainder theorem can be generalized to any ring, by using coprime ideals (also called comaximal ideals). Two ideals *I* and *J* are coprime if there are elements $i\in I$ and $j\in J$ such that $i+j=1$. This relation plays the role of Bézout's identity in the proofs related to this generalization, which, otherwise are very similar. The generalization may be stated as follows.

Let I_1, ..., I_k be two-sided ideals of a ring R that are pairwise coprime, and I be their intersection. Then we have the isomorphism,

$$R/I \to R/I_1 \times \cdots \times R/I_k$$
$$x \bmod I \mapsto (x \bmod I_1, \cdots, x \bmod I_k),$$

between the quotient ring $R >$ and the direct product of the R/I_i, where "$x \bmod I$" denotes the image of the element x in the quotient ring defined by the ideal I. Moreover, if R is commutative, then the ideal intersection I is equal to the product of the ideals R/I_i :

$$I = I_1 \cap I_2 \cap \ldots \cap I_k = I_1 I_2 \cdots I_k$$

Applications

Sequence Numbering

The Chinese remainder theorem has been used to construct a Gödel numbering for sequences, which is involved in the proof of Gödel's incompleteness theorems.

Fast Fourier Transform

The prime-factor FFT algorithm (also called Good-Thomas algorithm) uses the Chinese remainder theorem for reducing the computation of a fast Fourier transform of size $n_1 n_2$ to the computation of two fast Fourier transforms of smaller sizes n_1 and n_2 (providing that n_1 and n_2 are coprime).

Encryption

Most implementations of RSA use the Chinese remainder theorem during signing of HTTPS certificates and during decryption.

The Chinese remainder theorem can also be used in secret sharing, which consists of distributing a set of shares among a group of people who, all together (but no one alone), can recover a certain secret from the given set of shares. Each of the shares is represented in a congruence, and the solution of the system of congruences using the Chinese remainder theorem is the secret to be recovered. Secret sharing using the Chinese remainder theorem uses, along with the Chinese remainder theorem, special sequences of integers that guarantee the impossibility of recovering the secret from a set of shares with less than a certain cardinality.

Range Ambiguity Resolution

The range ambiguity resolution techniques used with medium pulse repetition frequency radar can be seen as a special case of the Chinese remainder theorem.

Dedekind's Theorem

Dedekind's Theorem on the Linear Independence of Characters. Let M be a monoid and k an integral domain, viewed as a monoid by considering the multiplication on k. Then any finite family $(f_i)_{i\in I}$ of distinct monoid homomorphisms $f_i : M \to k$ is linearly independent. In other words, every family $(\alpha_i)_{i\in I}$ of elements $\alpha_i \in k$ satisfying

$$\sum_{i\in I} \alpha_i f_i = 0$$

must be equal to the family $(0)_{i\in I}$.

Proof. First assume that k is a field, otherwise, replace the integral domain k by its quotient field, and nothing will change. We can linearly extend the monoid homomorphisms $f_i : M \to k$ to k-algebra homomorphisms $F_i : k[M] \to k$, where $k[M]$ is the monoid ring of M over k. Then, by linearity, the condition,

$$\sum_{i\in I} \alpha_i f_i = 0,$$

yields

$$\sum_{i\in I} \alpha_i F_i = 0.$$

Next, for $i, j \in I$; $i \neq j$ the two k-linear maps $F_i : k[M] \to k$ and $F_j : k[M] \to k$ are not proportional to each other. Otherwise f_i and f_j would also be proportional, and thus equal since as monoid homomorphisms they satisfy: $f_i(1) = 1 = f_j(1)$, which contradicts the assumption that they are distinct.

Therefore, the kernels Ker F_i and Ker F_j are distinct. Since $k[M]/\text{Ker } F_i \cong F_i(k[M]) = k$ is a field, Ker F_i is a maximal ideal of $k[M]$ for every $i \in I$. Because they are distinct and maximal the ideals Ker F_i and Ker F_j are coprime whenever $i \neq j$. The Chinese Remainder Theorem (for general rings) yields an isomorphism:

$$\phi : k[M]/K \to \prod_{i\in I} k[M]/\text{Ker}F$$

$$_i\phi(x+K) = (x + \text{Ker}F_i)_{i\in I}$$

where

$$K = \prod_{i\in I} \text{Ker}F_i = \bigcap_{i\in I} \text{Ker}F_i.$$

Consequently, the map

$$\Phi : k[M] \to \prod_{i\in I} k[M]/\text{Ker}F_i$$

$$\Phi(x) = (x + \text{Ker}F_i)_{i\in I}$$

is surjective. Under the isomorphisms $k[M]/\text{Ker } F_i \to F_i(k[M]) = k$, the map Φ corresponds to;

$$\psi : k[M] \to \prod_{i\in I} k$$
$$\psi(x) = [F_i(x)]_{i\in I}$$

Now,

$$\sum_{i\in I} \alpha_i F_i = 0$$

yields,

$$\sum_{i\in I} \alpha_i u_i = 0$$

for every vector $(u_i)_{i\in I}$ in the image of the map ψ. Since ψ is surjective, this means that,

$$\sum_{i\in I} \alpha_i u_i = 0$$

for every vector,

$$(u_i)_{i\in I} \in \prod_{i\in I} k.$$

Consequently, $(\alpha_i)_{i\in I} = (0)_{i\in I}$. QED.

Pentagonal Number Theorem

In mathematics, the pentagonal number theorem, originally due to Euler, relates the product and series representations of the Euler function. It states that

$$\prod_{n=1}^{\infty}\left(1-x^n\right) = \sum_{k=-\infty}^{\infty}\left(-1\right)^k x^{\frac{k}{2}(3k-1)} = 1+\sum_{k=1}^{\infty}(-1)^k\left(x^{k(3k+1)/2}+x^{k(3k-1)/2}\right).$$

In other words,

$$(1-x)(1-x^2)(1-x^3)\cdots = 1-x-x^2+x^5+x^7-x^{12}-x^{15}+x^{22}+x^{26}-\cdots.$$

The exponents 1, 2, 5, 7, 12, ... on the right hand side are given by the formula $g_k = k(3k-1)/2$ for $k = 1, -1, 2, -2, 3, \ldots$ and are called (generalized) pentagonal numbers. This holds as an identity of convergent power series for $|x|<1$, and also as an identity of formal power series.

A striking feature of this formula is the amount of cancellation in the expansion of the product.

Relation with Partitions

The identity implies a marvelous recurrence for calculating , the number of partitions of n;

$$p(n) = p(n-1) + p(n-2) - p(n-5) - p(n-7) + \cdots$$

or more formally,

$$p(n) = \sum_{k} (-1)^{k-1} p(n - g_k)$$

where the summation is over all nonzero integers k (positive and negative) and g_k is the kth pentagonal number.

Bijective Proof

The theorem can be given a combinatorial interpretation in terms of partitions. In particular, the left hand side is a generating function for the number of partitions of n into an even number of distinct parts minus the number of partitions of n into an odd number of distinct parts. Each partition of n into an even number of distinct parts contributes +1 to the coefficient of x^n; each partition into an odd number of distinct parts contributes −1. (The article on unrestricted partition functions discusses this type of generating function.)

For example, the coefficient of x^5 is +1 because there are two ways to split 5 into an even number of distinct parts (4+1 and 3+2), but only one way to do so for an odd number of distinct parts (the one-part partition 5). However, the coefficient of x^{12} is −1 because there are seven ways to partition 12 into an even number of distinct parts, but there are eight ways to partition 12 into an odd number of distinct parts.

This interpretation leads to a proof of the identity via involution (i.e. a bijection which is its own inverse). Consider the Ferrers diagram of any partition of n into distinct parts. For example, the diagram below shows n = 20 and the partition 20 = 7 + 6 + 4 + 3.

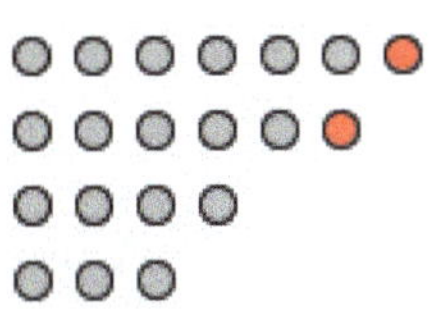

Let m be the number of elements in the smallest row of the diagram (m = 3 in the above example). Let s be the number of elements in the rightmost 45 degree line of the diagram (s = 2 dots in red above, since 7−1 = 6, but 6−1 > 4). If $m > s$, take the rightmost 45-degree line and move it to form a new row, as in the diagram below.

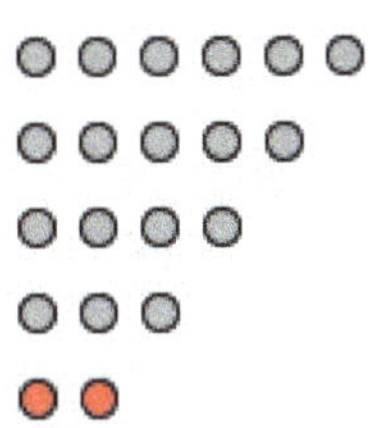

If m ≤ s (as in our newly formed diagram where m = 2, s = 5) we may reverse the process by moving the bottom row to form a new 45 degree line (adding 1 element to each of the first m rows), taking us back to the first diagram.

A bit of thought shows that this process always changes the parity of the number of rows, and applying the process twice brings us back to the original diagram. This enables us to pair off Ferrers diagrams contributing 1 and −1 to the x^n term of the series, resulting in a net coefficient of 0. This holds for every term *except* when the process cannot be performed on every Ferrers diagram with n dots. There are two such cases:

1) $m = s$ and the rightmost diagonal and bottom row meet. For example,

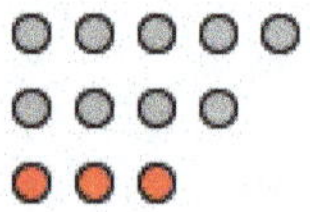

Attempting to perform the operation would lead us to:

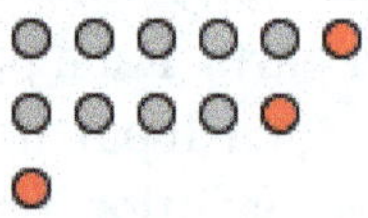

which fails to change the parity of the number of rows, and is not reversible in the sense that performing the operation again does *not* take us back to the original diagram. If there are m elements in the last row of the original diagram, then

$$n = m + (m+1) + (m+2) + \cdots + (2m-1) = \frac{m(3m-1)}{2} = \frac{k(3k-1)}{2}$$

where the new index k is taken to equal m. Note that the sign associated with this partition is $(-1)^s$, which by construction equals $(-1)^m$ and $(-1)^k$.

2) $m = s+1$ and the rightmost diagonal and bottom row meet. For example,

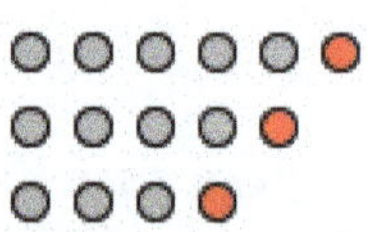

Our operation requires us to move the right diagonal to the bottom row, but that would lead to two rows of three elements, forbidden since we're counting partitions into distinct parts. This is the previous case but with one fewer row, so

$$n = m + (m+1) + (m+2) + \cdots + (2m-2) = \frac{(m-1)(3m-2)}{2} = \frac{k(3k-1)}{2},$$

where we take $k = 1-m$ (a negative integer). Here the associated sign is $(-1)^s$ with $s = m-1 = -k$, therefore the sign is again $(-1)^k$.

In summary, it has been shown that partitions into an even number of distinct parts and an odd number of distinct parts exactly cancel each other, except if n is a generalized pentagonal number $g_k = k(3k-1) >$, in which case there is exactly one Ferrers diagram left over. But this is precisely what the right side of the identity says should happen, so we are finished.

Partition Recurrence

We can rephrase the above proof, using partitions, which we denote as: $n = \lambda_1 + \lambda_2 + \cdots + \lambda_\ell$, where $\lambda_1 \geq \lambda_2 \geq \ldots \geq \lambda_\ell > 0$. The number of partitions of *n* is the partition function *p*(*n*) having generating function:

$$\sum_{n \in \mathbb{N}_0} p(n)x^n = \prod_{k \in \mathbb{N}} (1 - x^k)^{-1}$$

Note that is the reciprocal of the product on the left hand side of our identity:

$$\left(\sum_{n=0}^{\infty} p(n)x^n\right) \cdot \left(\prod_{n=1}^{\infty} (1 - x^n)\right) = 1$$

Let us denote the expansion of our product by $\prod_{n=1}^{\infty} (1 - x^n) = \sum_{n=0}^{\infty} a_n x^n$, so that

$$\left(\sum_{n=0}^{\infty} p(n)x^n\right) \cdot \left(\sum_{n=0}^{\infty} a_n x^n\right) = 1.$$

Multiplying out the left hand side and equating coefficients on the two sides, we obtain $a_o\, p(0) = 1$ and $\sum_{i=0}^{n} p(n-i)a_i = 0$ for all $n \geq 1$. This gives a recurrence relation defining *p*(*n*) in terms of a_n, and vice versa a recurrence for a_n in terms of *p*(*n*). Thus, our desired result:

$$a_i := \begin{cases} 1 & \text{if } i = \frac{1}{2}(3k^2 \pm k) \text{ and } k \text{ is even} \\ -1 & \text{if } i = \frac{1}{2}(3k^2 \pm k) \text{ and } k \text{ is odd} \\ 0 & \text{otherwise} \end{cases}$$

for $i \geq 1$ is equivalent to the identity $\sum_i (-1)^i p(n - g_i) = 0$, where $g_i := \frac{1}{2}(3i^2 - i)$ and *i* ranges over all integers such that $g_i := \frac{1}{2}(3i_2 - i)$ (this range includes both positive and negative i, so as to use both kinds of generalized pentagonal numbers). This in turn means:

$$\sum_{i \text{ even}} p(n - g_i) = \sum_{i \text{ odd}} p(n - g_i),.$$

In terms of sets of partitions, this is equivalent to saying that the following sets are of equal cardinality:

$$\mathcal{X} := \bigcup_{i \text{ even}} \mathcal{P}(n - g_i) \qquad \text{and} \qquad \mathcal{Y} := \bigcup_{i \text{ odd}} \mathcal{P}(n - g_i),$$

where $\mathcal{P}(n)$ denotes the set of all partitions of n. All that remains is to give a bijection from one set to the other, which is accomplished by the function φ from X to Y which maps the partition $\mathcal{P}(n-g_i) \ni \lambda : n-g_i = \lambda_1 + \lambda_2 + \cdots + \lambda_\ell$ to the partition $\lambda' = \varphi(\lambda)$ defined by:

$$\varphi(\lambda) := \begin{cases} \lambda' : n - g_{i-1} = (\ell + 3i - 2) + (\lambda_1 - 1) + \cdots + (\lambda_\ell - 1) & \text{if } \ell + 3i > \lambda_1 \\ \lambda' : n - g_{i+1} = (\lambda_2 + 1) + \cdots + (\lambda_\ell + 1) + \underbrace{1 + \cdots + 1}_{\lambda_1 - \ell - 3i} & \text{if } \ell + 3i \le \lambda_1 \end{cases}.$$

This is an involution (a self-inverse mapping), and thus in particular a bijection, which proves our claim and the identity.

Six Exponentials Theorem

In mathematics, specifically transcendental number theory, the six exponentials theorem is a result that, given the right conditions on the exponents, guarantees the transcendence of at least one of a set of exponentials.

Statement

If $x_1, x_2, \ldots, x_d$ are d complex numbers that are linearly independent over the rational numbers, and $y_1, y_2, \ldots, y_l$ are l complex numbers that are also linearly independent over the rational numbers, and if $dl > d + l$, then at least one of the following dl numbers is transcendental:

$$\exp(x_i y_j), \quad (1 \le i \le d, 1 \le j \le l).$$

The most interesting case is when $d = 3$ and $l = 2$, in which case there are six exponentials, hence the name of the result. The theorem is weaker than the related but thus far unproved four exponentials conjecture, whereby the strict inequality $dl > d + l$ is replaced with $dl \ge d + l$, thus allowing $d = l = 2$.

The theorem can be stated in terms of logarithms by introducing the set L of logarithms of algebraic numbers:

$$\mathcal{L} = \{\lambda \in \mathbb{C} : e^\lambda \in \overline{\mathbb{Q}}\}.$$

The theorem then says that if λ_{ij} are elements of L for $i = 1, 2$ and $j = 1, 2, 3$, such that λ_{11}, λ_{12}, and λ_{13} are linearly independent over the rational numbers, and λ_{11} and λ_{21} are also linearly independent over the rational numbers, then the matrix,

$$M = \begin{pmatrix} \lambda_{11} & \lambda_{12} & \lambda_{13} \\ \lambda_{21} & \lambda_{22} & \lambda_{23} \end{pmatrix}$$

has rank 2.

History

A special case of the result where x_1, x_2, and x_3 are logarithms of positive integers, $y_1 = 1$, and y_2 is real, was first mentioned in a paper by Leonidas Alaoglu and Paul Erdős from 1944 in which they try to prove that the ratio of consecutive colossally abundant numbers is always prime. They claimed that Carl Ludwig Siegel knew of a proof of this special case, but it is not recorded. Using the special case they manage to prove that the ratio of consecutive colossally abundant numbers is always either a prime or a semiprime.

The theorem was first explicitly stated and proved in its complete form independently by Serge Lang and Kanakanahalli Ramachandra in the 1960s.

Five Exponentials Theorem

A stronger, related result is the five exponentials theorem, which is as follows. Let x_1, x_2 and y_1, y_2 be two pairs of complex numbers, with each pair being linearly independent over the rational numbers, and let γ be a non-zero algebraic number. Then at least one of the following five numbers is transcendental:

$$e^{x_1 y_1}, e^{x_1 y_2}, e^{x_2 y_1}, e^{x_2 y_2}, e^{\gamma x_2 / x_1}.$$

This theorem implies the six exponentials theorem and in turn is implied by the as yet unproven four exponentials conjecture, which says that in fact one of the first four numbers on this list must be transcendental.

Sharp Six Exponentials Theorem

Another related result that implies both the six exponentials theorem and the five exponentials theorem is the sharp six exponentials theorem. This theorem is as follows. Let x_1, x_2, and x_3 be complex numbers that are linearly independent over the rational numbers, and let y_1 and y_2 be a pair of complex numbers that are linearly independent over the rational numbers, and suppose that β_{ij} are six algebraic numbers for $1 \leq i \leq 3$ and $1 \leq j \leq 2$ such that the following six numbers are algebraic:

$$e^{x_1 y_1 - \beta_{11}}, e^{x_1 y_2 - \beta_{12}}, e^{x_2 y_1 - \beta_{21}}, e^{x_2 y_2 - \beta_{22}}, e^{x_3 y_1 - \beta_{31}}, e^{x_3 y_2 - \beta_{32}}.$$

Then $x_i y_j = \beta_{ij}$ for $1 \leq i \leq 3$ and $1 \leq j \leq 2$. The six exponentials theorem then follows by setting $\beta_{ij} = 0$

for every i and j, while the five exponentials theorem follows by setting $x_3 = \gamma/x_1$ and using Baker's theorem to ensure that the x_i are linearly independent.

There is a sharp version of the five exponentials theorem as well, although it as yet unproven so is known as the sharp five exponentials conjecture. This conjecture implies both the sharp six exponentials theorem and the five exponentials theorem, and is stated as follows. Let x_1, x_2 and y_1, y_2 be two pairs of complex numbers, with each pair being linearly independent over the rational numbers, and let α, β_{11}, β_{12}, β_{21}, β_{22}, and γ be six algebraic numbers with $\gamma \neq 0$ such that the following five numbers are algebraic:

$$e^{x_1 y_1 - \beta_{11}}, e^{x_1 y_2 - \beta_{12}}, e^{x_2 y_1 - \beta_{21}}, e^{x_2 y_2 - \beta_{22}}, e^{(\gamma x_2/x_1) - \alpha}.$$

Then $x_i y_j = \beta_{ij}$ for $1 \leq i, j \leq 2$ and $\gamma x_2 = \alpha x_1$.

A consequence of this conjecture that isn't currently known would be the transcendence of e^{π^2}, by setting $x_1 = y_1 = \beta_{11} = 1$, $x_2 = y_2 = i\pi$, and all the other values in the statement to be zero.

Strong Six Exponentials Theorem

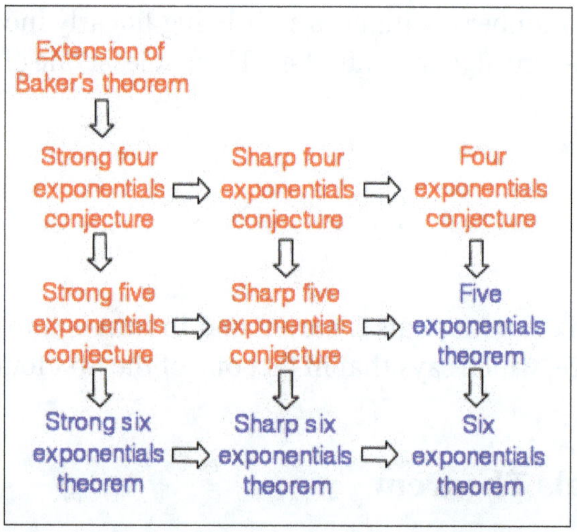

The logical implications between the various problems in this circle. Those in red are as yet unproven while those in blue are known results. The top most result refers to that discussed at Baker's theorem, while the four exponentials conjectures are detailed at the four exponentials conjecture article.

A further strengthening of the theorems and conjectures in this area are the strong versions. The strong six exponentials theorem is a result proved by Damien Roy that implies the sharp six exponentials theorem. This result concerns the vector space over the algebraic numbers generated by 1 and all logarithms of algebraic numbers, denoted here as *L*□. So *L*□ is the set of all complex numbers of the form

$$\beta_0 + \sum_{i=1}^{n} \beta_i \log \alpha_i,$$

for some $n \geq 0$, where all the β_i and α_i are algebraic and every branch of the logarithm is considered. The strong six exponentials theorem then says that if x_1, x_2, and x_3 are complex numbers that are linearly independent over the algebraic numbers, and if y_1 and y_2 are a pair of complex numbers that are also linearly independent over the algebraic numbers then at least one of the six numbers $x_i y_j$ for $1 \leq i \leq 3$ and $1 \leq j \leq 2$ is not in L^*. This is stronger than the standard six exponentials theorem which says that one of these six numbers is not simply the logarithm of an algebraic number.

There is also a strong five exponentials conjecture formulated by Michel Waldschmidt It would imply both, the strong six exponentials theorem and the sharp five exponentials conjecture. This conjecture claims that if x_1, x_2 and y_1, y_2 are two pairs of complex numbers, with each pair being linearly independent over the algebraic numbers, then at least one of the following five numbers is not in L^*:

$$x_1 y_1, x_1 y_2, x_2 y_1, x_2 y_2, x_1 >$$

All the above conjectures and theorems are consequences of the unproven extension of Baker's theorem, that logarithms of algebraic numbers that are linearly independent over the rational numbers are automatically algebraically independent too. The diagram on the right shows the logical implications between all these results.

Generalization to Commutative Group Varieties

The exponential function e^z uniformizes the exponential map of the multiplicative group G_m. Therefore, we can reformulate the six exponential theorem more abstractly as follows:

> Let $G = G_m \times G_m$ and take $u : C \to G(C)$ to be a non-zero complex-analytic group homomorphism. Define L to be the set of complex numbers l for which $u(l)$ is an algebraic point of G. If a minimal generating set of L over Q has more than two elements then the image $u(C)$ is an algebraic subgroup of $G(C)$.

(In order to derive the classical statement, set $u(z) = (e^{y_1 z}; e^{y_2 z})$ and note that $Qx_1 + Qx_2 + Qx_3$ is a subset of L).

In this way, the statement of the six exponentials theorem can be generalized to an arbitrary commutative group variety G over the field of algebraic numbers. This generalized six exponential conjecture, however, seems out of scope at the current state of transcendental number theory.

For the special, but interesting cases $G = G_m \times E$ and $G = E \times E'$, where E, E' are elliptic curves over the field of algebraic numbers, results towards the generalized six exponential conjecture were proven by Aleksander Momot. These results involve the exponential function e^z and a Weierstrass function $\wp$ resp. two Weierstrass functions $\wp, \wp'$ with algebraic invariants $g_2, g_3, g_{2'}, g_{3'}$, instead of the two exponential functions $e^{y_1 z}, e^{y_2 z}$ in the classical statement.

Let $G = G_m \times E$ and suppose E is not isogenous to a curve over a real field and that $u(C)$ is not an algebraic subgroup of $G(C)$. Then L is generated over Q either by two elements x_1, x_2, or three elements x_1, x_2, x_3 which are not all contained in a real line Rc, where c is a non-zero complex number. A similar result is shown for $G = E \times E'$.

References

- Kanemitsu, Shigeru; Chaohua Jia (2002). Number theoretic methods: future trends. Springer. pp. 271–274. ISBN 978-1-4020-1080-4.
- Michel, Anthony N.; Herget, Charles J. (1993), Applied Algebra and Functional Analysis, New York: Dover Publications, ISBN 978-0-486-67598-5
- Baker, Alan (1990), Transcendental number theory, Cambridge Mathematical Library (2nd ed.), Cambridge University Press, ISBN 978-0-521-39791-9, MR 0422171
- Davenport, Harold (2000), Multiplicative number theory, Graduate Texts in Mathematics, 74 (3rd revised ed.), New York: Springer-Verlag, ISBN 978-0-387-95097-6, MR 1790423
- Tenenbaum, Gérald (1995), Introduction to Analytic and Probabilistic Number Theory, Cambridge studies in advanced mathematics, 46, Cambridge University Press, ISBN 0-521-41261-7
- "The Unreasonable Effectiveness of Number Theory", Stefan Andrus Burr, George E. Andrews, American Mathematical Soc., 1992, ISBN 978-0-8218-5501-0
- "Applications of number theory to numerical analysis", Lo-keng Hua, Luogeng Hua, Yuan Wang, Springer-Verlag, 1981, ISBN 978-3-540-10382-0
- Ireland, Kenneth; Rosen, Michael (1990), A Classical Introduction to Modern Number Theory (2nd ed.), Springer-Verlag, ISBN 0-387-97329-X

5

Combinatorics: A Comprehensive Study

The area of study within mathematics which deals with counting as well as some properties of finite structures is called combinatorics. Generating function is a method of encoding an infinite series of numbers by treating them as coefficients of a formal power series. This chapter discusses in detail the different aspects of combinatorics and generating function such as pigeonhole principle and inclusion-exclusion principle.

Pigeonhole Principle

Pigeons in holes. Here there are $n = 10$ pigeons in $m = 9$ holes. Since 10 is greater than 9, the pigeonhole principle says that at least one hole has more than one pigeon.

In mathematics, the pigeonhole principle states that if n items are put into m containers, with $n > m > 0$, then at least one container must contain more than one item. This theorem is exemplified in real life by truisms like "there must be at least two left gloves or two right gloves in a group of three gloves". It is an example of a counting argument, and despite seeming intuitive it can be used to demonstrate possibly unexpected results; for example, that two people in London have the same number of hairs on their heads.

The first formalization of the idea is believed to have been made by Peter Gustav Lejeune Dirichlet in 1834 under the name *Schubfachprinzip* ("drawer principle" or "shelf principle"). For this reason it is also commonly called Dirichlet's box principle or Dirichlet's drawer principle. This should not be confused with Dirichlet's principle, a term introduced by Riemann that refers to the minimum principle for harmonic functions.

The principle has several generalizations and can be stated in various ways. In a more quantified version: for natural numbers k and m, if $n = km + 1$ objects are distributed among m sets, then the pigeonhole principle asserts that at least one of the sets will contain at least $k + 1$ objects. For arbitrary n and m this generalizes to $k + 1 = \lfloor (n - 1)/m \rfloor + 1$, where $\lfloor \ldots \rfloor$ is the floor function.

Though the most straightforward application is to finite sets (such as pigeons and boxes), it is also used with infinite sets that cannot be put into one-to-one correspondence. To do so requires the formal statement of the pigeonhole principle, which is *"there does not exist an injective function whose codomain is smaller than its domain"*. Advanced mathematical proofs like Siegel's lemma build upon this more general concept.

Etymology

P.G.L. Dirichlet published his works in both French and German. The strict original meaning of either the German *Schubfach*, or the French *tiroir*, corresponds to the English *drawer*, an *open-topped box that can be slid in and out of the cabinet that contains it*. These terms were morphed to the word *pigeonhole*, standing for a *small open space in a desk, cabinet, or wall for keeping letters or papers*, metaphorically rooted in the structures that house pigeons. Considering the fact that Dirichlet's father was a postmaster, necessarily best acquainted to furniture of type *pigeonhole*, common for sorting letters in his business, the translation by *pigeonholes* may be a perfect transfer of Dirichlet's terms of understanding. The meaning, referring to some furniture features, has since been strongly overtaken and is fading, especially among those who do not speak English natively, but as a lingua franca in the scientific world, in favour of the more pictorial interpretation, literally involving pigeons and holes. It is interesting to note that the suggestive, though not misleading interpretation of "pigeonhole" as "dovecote" has lately found its way back to a German "re-translation" of the "pigeonhole"-principle as the "Taubenschlag"-principle.

Examples

Sock-picking

Assume a drawer contains a mixture of black socks and blue socks, each of which can be worn on either foot, and that you are pulling a number of socks from the drawer without looking. What is the minimum number of pulled socks required to guarantee a pair of the same color? Using the pigeonhole principle, to have at least one pair of the same color ($m = 2$ holes, one per color) using one pigeonhole per color, you need to pull only three socks from the drawer ($n = 3$ items). Either you have *three* of one color, or, exclusively, *two* of one color and *one* of the other.

Hand-shaking

If there are n people who can shake hands with one another (where $n > 1$), the pigeonhole principle shows that there is always a pair of people who will shake hands with the same number of people. As the 'holes', or m, correspond to number of hands shaken, and each person can shake hands with anybody from 0 to $n - 1$ other people, this creates $n - 1$ possible holes. This is because either the '0' or the '$n - 1$' hole must be empty (if one person shakes hands with everybody, it's not possible to have another person who shakes hands with nobody; likewise, if one person shakes hands with no one there cannot be a person who shakes hands with everybody). This leaves n people to be placed in at most $n - 1$ non-empty holes, guaranteeing duplication.

Hair-counting

We can demonstrate there must be at least two people in London with the same number of hairs on their heads as follows. Since a typical human head has an average of around 150,000 hairs, it is reasonable to assume (as an upper bound) that no one has more than 1,000,000 hairs on their head (m = 1 million holes). There are more than 1,000,000 people in London (n is bigger than 1 million items). Assigning a pigeonhole to each number of hairs on a person's head, and assign people to pigeonholes according to the number of hairs on their head, there must be at least two people assigned to the same pigeonhole by the 1,000,001st assignment (because they have the same number of hairs on their heads) (or, $n > m$). For the average case (m = 150,000) with the constraint: fewest overlaps, there will be at most one person assigned to every pigeonhole and the 150,001st person assigned to the same pigeonhole as someone else. In the absence of this constraint, there may be empty pigeonholes because the "collision" happens before we get to the 150,001st person. The principle just proves the existence of an overlap; it says nothing of the number of overlaps (which falls under the subject of probability distribution).

There is a passing, satirical, allusion in English to this version of the principle in *A History of the Athenian Society*, prefixed to ""A Supplement to the Athenian Oracle: Being a Collection of the Remaining Questions and Answers in the Old Athenian Mercuries"", (Printed for Andrew Bell, London, 1710). It seems that the question *whether there were any two persons in the World that have an equal number of hairs on their head?* had been raised in *The Athenian Mercury* before 1704.

Perhaps the first written reference to the pigeonhole principle appears in 1622 in a short sentence of the Latin work *Selectæ Propositiones*, by the French Jesuit Jean Leurechon, where he wrote "It is necessary that two men have the same number of hairs, écus, or other things, as each other."

The Birthday Problem

The birthday problem asks, for a set of n randomly chosen people, what is the probability that some pair of them will have the same birthday? By the pigeonhole principle, if there are 367 people in the room, we know that there is at least one pair who share the same birthday, as there are only 366 possible birthdays to choose from (including February 29, if present). The birthday "paradox" refers to the result that even if the group is as small as 23 individuals, there will still be a pair of people with the same birthday with a 50% probability. While at first glance this may seem surprising, it intuitively makes sense when considering that a comparison will actually be made between every possible pair of people rather than fixing one individual and comparing them solely to the rest of the group.

Softball Team

Imagine seven people who want to play softball (n = 7 items), with a limitation of only four softball teams (m = 4 holes) to choose from. The pigeonhole principle tells us that they cannot all play for different teams; there must be at least one team featuring at least two of the seven players:

$$\left\lfloor \frac{n-1}{m} \right\rfloor + 1 = \left\lfloor \frac{7-1}{4} \right\rfloor + 1 = \left\lfloor \frac{6}{4} \right\rfloor + 1 = 1 + 1 = 2$$

Subset Sum

Any subset of size six from the set $S = \{1,2,3,\ldots,9\}$ must contain two elements whose sum is 10. The pigeonholes will be labelled by the two element subsets {1,9}, {2,8}, {3,7}, {4,6} and the singleton

{5}, five pigeonholes in all. When the six "pigeons" (elements of the size six subset) are placed into these pigeonholes, each pigeon going into the pigeonhole that has it contained in its label, at least one of the pigeonholes labelled with a two element subset will have two pigeons in it.

Uses and Applications

The pigeonhole principle arises in computer science. For example, collisions are inevitable in a hash table because the number of possible keys exceeds the number of indices in the array. A hashing algorithm, no matter how clever, cannot avoid these collisions.

The principle can be used to prove that any lossless compression algorithm, provided it makes some inputs smaller (as the name compression suggests), will also make some other inputs larger. Otherwise, the set of all input sequences up to a given length L could be mapped to the (much) smaller set of all sequences of length less than L without collisions (because the compression is lossless), a possibility which the pigeonhole principle excludes.

A notable problem in mathematical analysis is, for a fixed irrational number a, to show that the set $\{[na]: n \text{ is an integer}\}$ of fractional parts is dense in $[0, 1]$. One finds that it is not easy to explicitly find integers n, m such that $|na - ma| < e$, where $e > 0$ is a small positive number and a is some arbitrary irrational number. But if one takes M such that $1/M < e$, by the pigeonhole principle there must be $n_1, n_2 \in \{1, 2, ..., M + 1\}$ such that $n_1 a$ and $n_2 a$ are in the same integer subdivision of size $1/M$ (there are only M such subdivisions between consecutive integers). In particular, we can find n_1, n_2 such that $n_1 a$ is in $(p + k/M, p + (k + 1)/M)$, and $n_2 a$ is in $(q + k/M, q + (k + 1)/M)$, for some p, q integers and k in $\{0, 1, ..., M - 1\}$. We can then easily verify that $(n_2 - n_1)a$ is in $(q - p - 1/M, q - p + 1/M)$. This implies that $[na] < 1/M < e$, where $n = n_2 - n_1$ or $n = n_1 - n_2$. This shows that 0 is a limit point of $\{[na]\}$. We can then use this fact to prove the case for p in $(0, 1]$: find n such that $[na] < 1/M < e$; then if $p \in (0, 1/M]$, we are done. Otherwise $p \in (j/M, (j + 1)/M]$, and by setting $k = \sup\{r \in N : r[na] < j/M\}$, one obtains $|[(k + 1)na] - p| < 1/M < e$.

Variants occurring in well known proofs: In the proof of the pumping lemma for regular languages, a version that mixes finite and infinite sets is used: If infinitely many objects are placed in finitely many boxes, then there exist two objects that share a box. In Fisk's solution of the Art gallery problem a sort of converse is used: If n objects are placed in k boxes, then there is a box containing at most n/k objects.

Alternate Formulations

The following are alternate formulations of the pigeonhole principle.

1. If n objects are distributed over m places, and if $n > m$, then some place receives at least two objects.

2. (equivalent formulation of 1) If n objects are distributed over n places in such a way that no place receives more than one object, then each place receives exactly one object.

3. If n objects are distributed over m places, and if $n < m$, then some place receives no object.

4. (equivalent formulation of 3) If n objects are distributed over n places in such a way that no place receives no object, then each place receives exactly one object.

Strong Form

Let $q_1, q_2, ..., q_n$ be positive integers. If

$$q_1 + q_2 + \cdots + q_n - n + 1$$

objects are distributed into n boxes, then either the first box contains at least q_1 objects, or the second box contains at least q_2 objects, ..., or the nth box contains at least q_n objects.

The simple form is obtained from this by taking $q_1 = q_2 = ... = q_n = 2$, which gives $n + 1$ objects. Taking $q_1 = q_2 = ... = q_n = r$ gives the more quantified version of the principle, namely:

Let n and r be positive integers. If $n(r - 1) + 1$ objects are distributed into n boxes, then at least one of the boxes contains r or more of the objects.

This can also be stated as, if k discrete objects are to be allocated to n containers, then at least one container must hold at least $\lceil k / n \rceil$ objects, where $\lceil \ \rceil$ is the ceiling function, denoting the smallest integer larger than or equal to x. Similarly, at least one container must hold no more than $\lfloor k / n \rfloor$ objects, where $\lfloor x \rfloor$ is the floor function, denoting the largest integer smaller than or equal to x.

Generalizations of the Pigeonhole Principle

A probabilistic generalization of the pigeonhole principle states that if n pigeons are randomly put into m pigeonholes with uniform probability $1/m$, then at least one pigeonhole will hold more than one pigeon with probability,

$$1 - \frac{(m)_n}{m^n},$$

where $(m)_n$ is the falling factorial $m(m - 1)(m - 2)...(m - n + 1)$. For $n = 0$ and for $n = 1$ (and $m > 0$), that probability is zero; in other words, if there is just one pigeon, there cannot be a conflict. For $n > m$ (more pigeons than pigeonholes) it is one, in which case it coincides with the ordinary pigeonhole principle. But even if the number of pigeons does not exceed the number of pigeonholes ($n \leq m$), due to the random nature of the assignment of pigeons to pigeonholes there is often a substantial chance that clashes will occur. For example, if 2 pigeons are randomly assigned to 4 pigeonholes, there is a 25% chance that at least one pigeonhole will hold more than one pigeon; for 5 pigeons and 10 holes, that probability is 69.76%; and for 10 pigeons and 20 holes it is about 93.45%. If the number of holes stays fixed, there is always a greater probability of a pair when you add more pigeons. This problem is treated at much greater length in the birthday paradox.

A further probabilistic generalization is that when a real-valued random variable X has a finite mean $E(X)$, then the probability is nonzero that X is greater than or equal to $E(X)$, and similarly the probability is nonzero that X is less than or equal to $E(X)$. To see that this implies the standard pigeonhole principle, take any fixed arrangement of n pigeons into m holes and let X be the number of pigeons in a hole chosen uniformly at random. The mean of X is n/m, so if there are more pigeons than holes the mean is greater than one. Therefore, X is sometimes at least 2.

Infinite Sets

The pigeonhole principle can be extended to infinite sets by phrasing it in terms of cardinal numbers: if the cardinality of set A is greater than the cardinality of set B, then there is no injection from A to B. However, in this form the principle is tautological, since the meaning of the statement that the cardinality of set A is greater than the cardinality of set B is exactly that there is no injective map from A to B. However, adding at least one element to a finite set is sufficient to ensure that the cardinality increases.

Another way to phrase the pigeonhole principle for finite sets is similar to the principle that finite sets are Dedekind finite: Let A and B be finite sets. If there is a surjection from A to B that is not injective, then no surjection from A to B is injective. In fact no function of any kind from A to B is injective. This is not true for infinite sets: Consider the function on the natural numbers that sends 1 and 2 to 1, 3 and 4 to 2, 5 and 6 to 3, and so on.

There is a similar principle for infinite sets: If uncountably many pigeons are stuffed into countably many pigeonholes, there will exist at least one pigeonhole having uncountably many pigeons stuffed into it.

This principle is not a generalization of the pigeonhole principle for finite sets however: It is in general false for finite sets. In technical terms it says that if A and B are finite sets such that any surjective function from A to B is not injective, then there exists an element of b of B such that there exists a bijection between the preimage of b and A. This is a quite different statement, and is absurd for large finite cardinalities.

Quantum Mechanics

Yakir Aharonov et al. have presented arguments that the pigeonhole principle may be violated in quantum mechanics, and proposed interferometric experiments to test the pigeonhole principle in quantum mechanics.

The pigeonhole principle states that if there are n + 1 pigeons and n holes (boxes), then there is at least one hole (box) that contains two or more pigeons. It can be easily verified that the pigeonhole principle is equivalent to the following statements:

1. If m pigeons are put into m pigeonholes, there is an empty hole if and only if there's a hole with more than one pigeon.

2. If n pigeons are put into m pigeonholes, with n > m, then there is a hole with more than one pigeon.

3. For two finite sets A and B, there exists a one to one and onto function f : A → B if and only if |A| = |B|.

Remark: Recall that the expression $\lceil x \rceil$, called the ceiling function, is the smallest integer ℓ, such that $\ell \geq x$ and the expression $\lfloor x \rfloor$, called the floor function, is the largest integer k, such that $k \leq x$.

1.[Generalized Pigeonhole Principle] if there are n pigeons and m holes with n > m, then there is at least one hole that contains $\lceil \frac{n}{m} \rceil$ pigeons.

2. Dirichlet was the one who popularized this principle.

Example: Let a be an irrational number. Then prove that there exist infinitely many rational numbers $s = \frac{p}{q}$, such that $|a - s| < \frac{1}{q^2}$.

Proof. Let $N \in \mathbb{N}$. Without loss of generality, we assume that a > 0. By {α}, we will denote the fractional part of α. That is, {α} = α − ⌊α⌋.

Now, consider the fractional parts {0}, {a}, {2a}, . . . , {Na} of the first (N + 1) multiples of a and the N subintervals $\left[0, \frac{1}{N}\right), \left[\frac{1}{N}, \frac{2}{N}\right), \ldots, \left[\frac{N-1}{N}, 1\right)$ *of* $[0, 1)$. Clearly {ka}, for k a positive integer, cannot be an integer as a is an irrational number. Thus, by the pigeonhole principle, two of the above fractional parts must fall into the same subinterval. That is, there exist integers u, v and w such that u > v but,

$$\{ua\} \in \left[\frac{w}{N}, \frac{w+1}{N}\right) \text{ and } \{va\} \in \left[\frac{w}{N}, \frac{w+1}{N}\right).$$

Thus, $|\{ua\} - \{va\}| < \frac{1}{N}$ and $|\{ua\} - \{va\}| = |(u - v)a - (\lfloor ua \rfloor - \lfloor va \rfloor)|$. Now, let $q = u - v$ and $p = \lfloor ua \rfloor - \lfloor va \rfloor$. Then $p, q \in \mathbb{Z}, q \neq 0$ and $|qa - p| < \frac{1}{N}$. Dividing by q, we get

$$\left|a - \frac{p}{q}\right| < \frac{1}{Nq} \leq \frac{1}{q^2} \quad as \quad 0 < q \leq N.$$

Therefore, we have found a rational number $\frac{p}{q}$ such that $\left|a - \frac{p}{q}\right| < \frac{1}{q^2}$. We will now show that the number of such pairs (p, q) is infinite.

On the contrary, assume that there are only a finite number of rational numbers, say $r_1, r_2, \ldots, r_M$ such that

$$r_i = \frac{p_i}{q_i}, \text{ for } i = 1, \ldots, M \text{ and } |a - r_i| < \frac{1}{q_i^2}.$$

Since a is an irrational number, none of the differences $|a - r_i|$, for i = 1, 2, . . . , M, will be exactly 0. Therefore, there exists an integer Q such that,

$$|a - r_i| > \frac{1}{Q}, \quad \text{for all } i = 1, 2, \ldots, M.$$

We now, apply our earlier argument to this Q. The argument gives the existence of a fraction

$r = \frac{p}{q}$ such that $|a - r| > \frac{1}{Q_q} < \frac{1}{Q} < |a - r_i|$, *for* $1 \le \mathrm{i} \le \mathrm{M}$ Hence, $r \ne r_i$, for all i = 1, 2, . . . , M.

On the other hand, we also have, $|a - r| < \frac{1}{q^2}$ contradicting the assumption that the fractions r_i, for i = 1, 2, . . . , M, were all the fractions with this property.

Example: 1. Let $\{a_1, a_2, \ldots, a_{mn+1}\}$ be a sequence of distinct mn + 1 real numbers. Then prove that this sequence has a subsequence of either (m + 1) numbers that is strictly increasing or (n + 1) numbers that is strictly decreasing.

Observation: The statement is NOT TRUE if there are exactly mn numbers. For example, consider the sequence 4, 3, 2, 1, 8, 7, 6, 5, 12, 11, 10, 9 of 12 = 3 × 4 distinct numbers. This sequence neither has an increasing subsequence of 4 numbers nor a decreasing subsequence of 5 numbers. Also, observe that if we take any number different from 1, 2, . . . , 12 and place it at any position in the above sequence then it can be verified that there is either an increasing subsequence of 4 numbers or a decreasing subsequence of 5 numbers. For example, if we place 7.5,

(a) "before the number 4" or "between the numbers 8 and 7" or "after the number 9", then there is a decreasing subsequence of length 5.

(b) "between the numbers 4 and 8" or between the numbers 7 and 9, then there is an increasing subsequence of length 4.

Proof. Let T be the given sequence. That is, $T = \{a_k\}_{k=1}^{mn+1}$ and define $\ell_i = \max_s\{s :$ an increasing subsequence of length s exists starting with $a_i\}$.

Then there are mn+1 positive integers $\ell_1, \ell_2, \ldots, \ell_{mn+1}$. If there exists a j, $1 \le j \le mn+1$, such that $\ell_j \ge m + 1$, then by definition of ℓ_j, there exists an increasing sequence of length m + 1 starting with a_j and thus the result follows. So, on the contrary assume that $\ell_i \le m$, for $1 \le i \le mn + 1$.

That is, we have mn + 1 numbers $(\ell_1, \ldots, \ell_{mn+1})$ and all of them have to be put in the boxes numbered 1, 2, . . . , m. So, by the generalized pigeonhole principle, there are at least $\left\lceil \frac{mn+1}{m} \right\rceil = n+1$ numbers (ℓ_i's) that lies in the same box. Therefore, let us assume that there exist numbers $1 \le i_1 < i_2 < \cdots < i_{n+1} \le mn+1$, such that

$$\ell_{i_1} = \ell_{i_2} = \cdots = \ell_{i_{n+1}}.$$

That is, the length of the largest increasing subsequences of T starting with the numbers $a_{i_1}, a_{i_2}, \ldots, a_{i_{n+1}}$ are all equal. We now claim that $a_{i_1} > a_{i_2} > \ldots > a_{i_{n+1}}$.

We will show that $a_{i_1} > a_{i_2}$. A similar argument will give the other inequalities and hence the proof of the claim. On the contrary, let if $a_{i_1} < a_{i_2}$ (recall that a_i's are distinct) and consider a largest increasing subsequence $a_{i_2} = \alpha_1 < \alpha_2 < \cdots < \alpha_{\ell_{i_2}}$ of T, starting with a_{i_2}, that has length ℓ_{i_2}. This subsequence with the assumption that $a_{i_1} < a_{i_2}$ gives an increasing subsequence,

$$a_{i_1} < a_{i_2} = \alpha_1 < \alpha_2 < \cdots < \alpha_{li_2}$$

of T, starting with a_{i_1}, of length $\ell_{i_2} + 1$. So, by definition of ℓ_i's, $\ell_{i_1} \geq \ell_{i_2} + 1$. This gives a contradiction to the equality, $\ell_{i_1} \quad \ell_{i_2}$, in earlier Equation. Hence the proof of the example is complete.

2. Prove that there exist two powers of 3 whose difference is divisible by 2011.

Proof. Consider the set S = {1 = 3^0, 3, 3^2, 3^3, . . . , 3^{2011}}. Then |S| = 2012. Also, we know that when we divide positive integers by 2011 then the possible remainders are 0, 1, 2, . . . , 2010 (corresponding to exactly 2011 boxes). So, if we divide the numbers in S with 2011, then by pigeonhole principle there will exist at least two numbers 0 ≤ i < j ≤ 2011, such that the remainders of 3^j and 3^i, when divided by 2011, are equal. That is, 2011 divides $3^j - 3^i$. Hence, this completes the proof.

Observe that this argument also implies that "there exists a positive integer ℓ such that 2011 divides $3^\ell - 1$" or "there exists a positive power of 3 that leaves a remainder 1 when divided by 2011" as gcd(3, 2011) = 1.

3. Prove that there exists a power of three that ends with 0001.

Proof. Consider the set S = {1 = 3^0, 3, 3^2, 3^3, . . .}. Now, let us divide each element of S by 10^4. As |S| > 10^4, there exist i > j such that the remainders of 3^i and 3^j, when are divided by 10^4, are equal. But gcd(10^4, 3) = 1 and thus, 10^4 divides $3^\ell - 1$. That is, $3^\ell - 1 = s \bullet 10^4$ for some positive integer s. That is, $3^\ell = s \bullet 10^4 + 1$ and hence the result follows.

Inclusion–Exclusion Principle

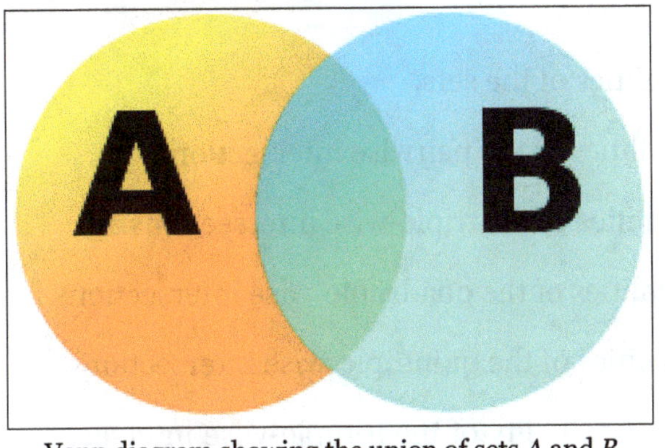

Venn diagram showing the union of sets *A* and *B*.

In combinatorics (combinatorial mathematics), the inclusion–exclusion principle is a counting technique which generalizes the familiar method of obtaining the number of elements in the union of two finite sets; symbolically expressed as,

$$|A \cup B| = |A| + |B| - |A \cap B|,$$

where *A* and *B* are two finite sets and |*S*| indicates the cardinality of a set *S* (which may be considered as the number of elements of the set, if the set is finite). The formula expresses the fact that the sum of the sizes of the two sets may be too large since some elements may be counted twice. The double-counted elements are those in the intersection of the two sets and the count is corrected by subtracting the size of the intersection.

The principle is more clearly seen in the case of three sets, which for the sets A, B and C is given by

$$|A \cup B \cup C| = |A| + |B| + |C| - |A \cap B| - |A \cap C| - |B \cap C| + |A \cap B \cap C|.$$

This formula can be verified by counting how many times each region in the Venn diagram figure is included in the right-hand side of the formula. In this case, when removing the contributions of over-counted elements, the number of elements in the mutual intersection of the three sets has been subtracted too often, so must be added back in to get the correct total.

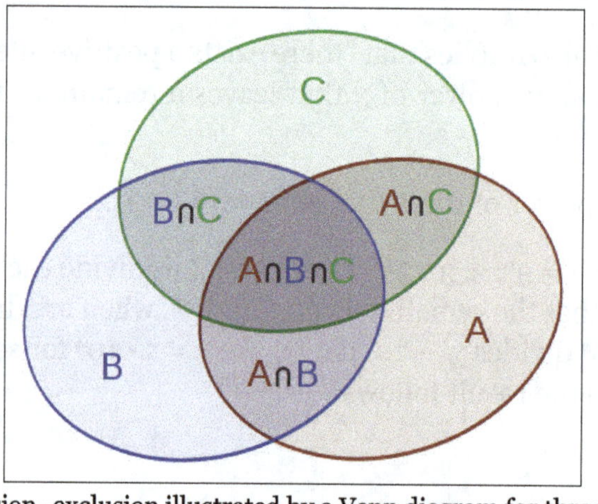

Inclusion–exclusion illustrated by a Venn diagram for three sets.

Generalizing the results of these examples gives the principle of inclusion–exclusion. To find the cardinality of the union of n sets:

1. Include the cardinalities of the sets.
2. Exclude the cardinalities of the pairwise intersections.
3. Include the cardinalities of the triple-wise intersections.
4. Exclude the cardinalities of the quadruple-wise intersections.
5. Include the cardinalities of the quintuple-wise intersections.
6. Continue, until the cardinality of the n-tuple-wise intersection is included (if n is odd) or excluded (n even).

The name comes from the idea that the principle is based on over-generous *inclusion*, followed by compensating *exclusion*. This concept is attributed to Abraham de Moivre (1718); but it first appears in a paper of Daniel da Silva (1854), and later in a paper by J. J. Sylvester (1883). Sometimes the principle is referred to as the formula of Da Silva, or Sylvester due to these publications. The principle is an example of the sieve method extensively used in number theory and is sometimes referred to as the *sieve formula*, though Legendre already used a similar device in a sieve context in 1808.

As finite probabilities are computed as counts relative to the cardinality of the probability space, the formulas for the principle of inclusion–exclusion remain valid when the cardinalities of the sets are replaced by finite probabilities. More generally, both versions of the principle can be put under the common umbrella of measure theory.

In a very abstract setting, the principle of inclusion–exclusion can be expressed as the calculation of the inverse of a certain matrix. This inverse has a special structure, making the principle an extremely valuable technique in combinatorics and related areas of mathematics. As Gian-Carlo Rota put it:

"One of the most useful principles of enumeration in discrete probability and combinatorial theory is the celebrated principle of inclusion–exclusion. When skillfully applied, this principle has yielded the solution to many a combinatorial problem."

Statement

In its general form, the principle of inclusion–exclusion states that for finite sets $A_1, ..., A_n$, one has the identity

$$\left|\bigcup_{i=1}^{n} A_i\right| = \sum_{i=1}^{n} |A_i| - \sum_{1\le i<j\le n} |A_i \cap A_j| + \cdots$$
$$\cdots + \sum_{1\le i<j<k\le n} |A_i \cap A_j \cap A_k| - \cdots + (-1)^{n-1} \left|A_1 \cap \cdots \cap A_n\right|.$$

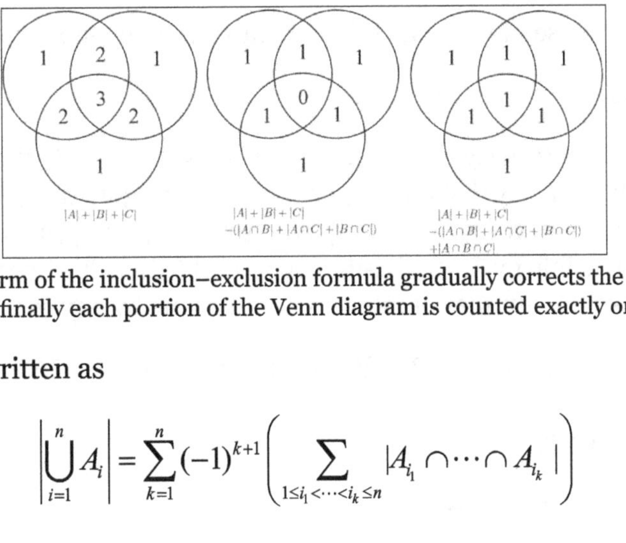

Each term of the inclusion–exclusion formula gradually corrects the count until finally each portion of the Venn diagram is counted exactly once.

This can be compactly written as

$$\left|\bigcup_{i=1}^{n} A_i\right| = \sum_{k=1}^{n} (-1)^{k+1} \left(\sum_{1\le i_1<\cdots<i_k\le n} |A_{i_1} \cap \cdots \cap A_{i_k}| \right)$$

or

$$\left|\bigcup_{i=1}^{n} A_i\right| = \sum_{\varnothing \ne J \subseteq \{1,2,\ldots,n\}} (-1)^{|J|-1} \left|\bigcap_{j\in J} A_j\right|.$$

In words, to count the number of elements in a finite union of finite sets, first sum the cardinalities of the individual sets, then subtract the number of elements which appear in more than one set, then add back the number of elements which appear in more than two sets, then subtract the number of elements which appear in more than three sets, and so on. This process naturally ends since there can be no elements which appear in more than the number of sets in the union.

In applications it is common to see the principle expressed in its complementary form. That is, letting S be a finite universal set containing all of the A_i and letting $\overline{A_i}$ denote the complement of A_i in S, by De Morgan's laws we have

$$\left|\bigcap_{i=1}^{n} \overline{A_i}\right| = \left|S - \bigcup_{i=1}^{n} A_i\right|$$

$$= |S| - \sum_{i=1}^{n} |A_i| + \sum_{1 \le i < j \le n} |A_i \cap A_j| - \cdots + (-1)^n |A_1 \cap \cdots \cap A_n|.$$

As another variant of the statement, let $P_1, P_2, \ldots, P_n$ be a list of properties that elements of a set S may or may not have, then the principle of inclusion–exclusion provides a way to calculate the number of elements of S which have none of the properties. Just let A_i be the subset of elements of S which have the property P_i and use the principle in its complementary form. This variant is due to J. J. Sylvester.

Examples

Counting Integers

As a simple example of the use of the principle of inclusion–exclusion, consider the question:

How many integers in {1,...,100} are not divisible by 2, 3 or 5?

Let $S = \{1,\ldots,100\}$ and P_1 the property that an integer is divisible by 2, P_2 the property that an integer is divisible by 3 and P_3 the property that an integer is divisible by 5. Letting A_i be the subset of S whose elements have property P_i we have by elementary counting: $|A_1| = 50$, $|A_2| = 33$, and $|A_3| = 20$. There are 16 of these integers divisible by 6, 10 divisible by 10 and 6 divisible by 15. Finally, there are just 3 integers divisible by 30, so the number of integers not divisible by any of 2, 3 or 5 is given by:

$$100 - (50 + 33 + 20) + (16 + 10 + 6) - 3 = 26.$$

Counting Derangements

A more complex example is the following:

Suppose there is a deck of n cards numbered from 1 to n. Suppose a card numbered m is in the correct position if it is the mth card in the deck. How many ways, W, can the cards be shuffled with at least 1 card being in the correct position?

Begin by defining set A_m, which is all of the orderings of cards with the mth card correct. Then the number of orders, W, with *at least* one card being in the correct position, m, is

$$W = \left|\bigcup_{m=1}^{n} A_m\right|.$$

Apply the principle of inclusion–exclusion,

$$W = \sum_{m_1=1}^{n} |A_{m_1}| - \sum_{1 \le m_1 < m_2 \le n} |A_{m_1} \cap A_{m_2}| + \cdots$$
$$\cdots + \sum_{1 \le m_1 < m_2 < m_3 \le n} |A_{m_1} \cap A_{m_2} \cap A_{m_3}| - \cdots$$
$$\cdots + (-1)^{p-1} \sum_{1 \le m_1 < \cdots < m_p \le n} |A_{m_1} \cap \cdots \cap A_{m_p}| \cdots.$$

Each value $A_{m_1} \cap \cdots \cap A_{m_p}$ represents the set of shuffles having at least p values m_1, ..., m_p in the correct position. Note that the number of shuffles with at least p values correct only depends on p, not on the particular values of m. For example, the number of shuffles having the 1st, 3rd, and 17th cards in the correct position is the same as the number of shuffles having the 2nd, 5th, and 13th cards in the correct positions. It only matters that of the n cards, 3 were chosen to be in the correct position. Thus there are $\binom{n}{p}$ equal terms in the pth summation.

$$W = \binom{n}{1} |A_1| - \binom{n}{2} |A_1 \cap A_2| + \binom{n}{3} |A_1 \cap A_2 \cap A_3| - \cdots$$
$$\cdots + (-1)^{p-1} \binom{n}{p} |A_1 \cap \cdots \cap A_p| \pm \cdots.$$

$|A_1 \cap \cdots \cap A_p|$ is the number of orderings having p elements in the correct position, which is equal to the number of ways of ordering the remaining $n - p$ elements, or $(n - p)!$. Thus we finally get:

$$W = \binom{n}{1}(n-1)! - \binom{n}{2}(n-2)! + \binom{n}{3}(n-3)! - \cdots$$
$$\cdots + (-1)^{p-1} \binom{n}{p}(n-p)! \pm \cdots.$$

$$W = \sum_{p=1}^{n} (-1)^{p-1} \binom{n}{p}(n-p)!.$$

Noting that $\binom{n}{p} = \frac{n!}{p!(n-p)!}$, this reduces to

$$W = \sum_{p=1}^{n} (-1)^{p-1} \frac{n!}{p!}.$$

A permutation where *no* card is in the correct position is called a derangement. Taking $n!$ to be the total number of permutations, the probability Q that a random shuffle produces a derangement is given by,

$$Q = 1 - \frac{W}{n!} = \sum_{p=0}^{n} \frac{(-1)^p}{p!},$$

a truncation to $n + 1$ terms of the Taylor expansion of e^{-1}. Thus the probability of guessing an order for a shuffled deck of cards and being incorrect about every card is approximately $1/e$ or 37%.

A Special Case

The situation that appears in the derangement example above occurs often enough to merit special attention. Namely, when the size of the intersection sets appearing in the formulas for the principle of inclusion–exclusion depend only on the number of sets in the intersections and not on which sets appear. More formally, if the intersection

$$A_J := \bigcap_{j \in J} A_j$$

has the same cardinality, say $\alpha_k = |A_J|$, for every k-element subset J of $\{1, ..., n\}$, then

$$\left| \bigcup_{i=1}^{n} A_i \right| = \sum_{k=1}^{n} (-1)^{k-1} \binom{n}{k} \alpha_k.$$

Or, in the complementary form, where the universal set S has cardinality α_0,

$$\left| S \setminus \bigcup_{i=1}^{n} A_i \right| = \sum_{k=0}^{n} (-1)^{k} \binom{n}{k} \alpha_k.$$

A Generalization

Given a family (repeats allowed) of subsets $A_1, A_2, ..., A_n$ of a universal set S, the principle of inclusion–exclusion calculates the number of elements of S in none of these subsets. A generalization of this concept would calculate the number of elements of S which appear in exactly some fixed m of these sets.

Let $N = [n] = \{1,2,...,n\}$. If we define $A_\varnothing = S$, then the principle of inclusion–exclusion can be written as, using the notation of the previous section; the number of elements of S contained in none of the A_i is:

$$\sum_{J \subseteq [n]} (-1)^{|J|} \, | A_J |.,$$

If I is a fixed subset of the index set N, then the number of elements which belong to A_i for all i in I and for no other values is:

$$\sum_{I \subseteq J} (-1)^{|J|-|I|} \, | A_J |.$$

Define the sets

$$B_k = A_{I \cup \{k\}} \text{ for } k \in N \setminus I.$$

We seek the number of elements in none of the B_k which, by the principle of inclusion–exclusion (with $B_\varnothing = A_I$), is

$$\sum_{K \subseteq N \setminus I} (-1)^{|K|} \, |B_K| .$$

The correspondence $K \leftrightarrow J = I \cup K$ between subsets of $N \setminus I$ and subsets of N containing I is a bijection and if J and K correspond under this map then $B_K = A_J$, showing that the result is valid.

In Probability

In probability, for events A_1, ..., A_n in a probability space $(\Omega, \mathcal{F}, \mathbb{P})$, , the inclusion–exclusion principle becomes for $n = 2$

$$\mathbb{P}(A_1 \cup A_2) = \mathbb{P}(A_1) + \mathbb{P}(A_2) - \mathbb{P}(A_1 \cap A_2),$$

for $n = 3$

$$\mathbb{P}(A_1 \cup A_2 \cup A_3) = \mathbb{P}(A_1) + \mathbb{P}(A_2) + \mathbb{P}(A_3) \quad - \mathbb{P}(A_1 \cap A_2) - \mathbb{P}(A_1 \cap A_3) - \mathbb{P}(A_2 \cap A_3) \quad + \mathbb{P}(A_1 \cap A_2 \cap A_3)$$

and in general,

$$\mathbb{P}\Big(\bigcup_{i=1}^{n} A_i\Big) = \sum_{i=1}^{n} \mathbb{P}(A_i) - \sum_{i<j} \mathbb{P}(A_i \cap A_j) \quad + \sum_{i<j<k} \mathbb{P}(A_i \cap A_j \cap A_k) - \cdots + (-1)^{n-1} \mathbb{P}\Big(\bigcap_{i=1}^{n} A_i\Big),$$

which can be written in closed form as,

$$\mathbb{P}\Big(\bigcup_{i=1}^{n} A_i\Big) = \sum_{k=1}^{n} \left((-1)^{k-1} \sum_{\substack{I \subset \{1,\ldots,n\} \\ |I|=k}} \mathbb{P}(A_I) \right),$$

where the last sum runs over all subsets I of the indices 1, ..., n which contain exactly k elements, and

$$A_I := \bigcap_{i \in I} A_i$$

denotes the intersection of all those A_i with index in I.

According to the Bonferroni inequalities, the sum of the first terms in the formula is alternately an upper bound and a lower bound for the LHS. This can be used in cases where the full formula is too cumbersome.

For a general measure space (S,Σ,μ) and measurable subsets A_1, ..., A_n of finite measure, the above identities also hold when the probability measure $\mathbb{P}$ is replaced by the measure μ.

Special Case

If, in the probabilistic version of the inclusion–exclusion principle, the probability of the intersection A_I only depends on the cardinality of I, meaning that for every k in {1, ..., n} there is an a_k such that

$$a_k = \mathbb{P}(A_I) \text{ for every } I \subset \{1,\ldots,n\} \text{ with } |I| = k,$$

then the above formula simplifies to,

$$\mathbb{P}\left(\bigcup_{i=1}^{n} A_i\right) = \sum_{k=1}^{n} (-1)^{k-1} \binom{n}{k} a_k$$

due to the combinatorial interpretation of the binomial coefficient $\binom{n}{k}$.

An analogous simplification is possible in the case of a general measure space (S,Σ,μ) and measurable subsets A_1, ..., A_n of finite measure.

Other Forms

The principle is sometimes stated in the form that says that if,

$$g(A) = \sum_{S:S \subseteq A} f(S)$$

then

$$f(A) = \sum_{S:S \subseteq A} (-1)^{|A|-|S|} g(S) \qquad (**)$$

The combinatorial and the probabilistic version of the inclusion–exclusion principle are instances of (**).

Proof

Take $\underline{m} = \{1,2,\ldots,m\}, f(\underline{m}) = 0$, and

$$f(S) = \left|\bigcap_{i \in \underline{m} \setminus S} A_i \setminus \bigcup_{i \in S} A_i\right| \text{ and } f(S) = \mathbb{P}\left(\bigcap_{i \in \underline{m} \setminus S} A_i \setminus \bigcup_{i \in S} A_i\right)$$

respectively for all sets S with $S \subsetneq \underline{m}$. Then we obtain,

$$g(A) = \left|\bigcap_{i \in \underline{m} \setminus A} A_i\right|, \; g(\underline{m}) = \left|\bigcup_{i \in \underline{m}} A_i\right| \text{ and } g(A) = \mathbb{P}\left(\bigcap_{i \in \underline{m} \setminus A} A_i\right), \; g(\underline{m}) = \mathbb{P}\left(\bigcup_{i \in \underline{m}} A_i\right)$$

respectively for all sets A with $A \subsetneq \underline{m}$. This is because elements a of $\bigcap_{i \in \underline{m} \setminus A} A_i$ can be contained in other A_i' s (A_i's with $i \in A$) as well, and the $\cap \setminus \cup$- formula runs exactly through all possible extensions of the sets $\{A_i \mid i \in \underline{m} \setminus A\}$ with other A_i's, counting a only for the set that matches the membership behavior of a, if S runs through all subsets of A (as in the definition of $g(A)$).

Since $f(\underline{m}) = 0$, we obtain from (**) with $A = \underline{m}$ that

$$\sum_{\underline{m} \supseteq T \supsetneq \varnothing} (-1)^{|T|-1} g(\underline{m} \setminus T) = \sum_{\varnothing \subseteq S \subsetneq \underline{m}} (-1)^{m-|S|-1} g(S) = g(\underline{m})$$

and by interchanging sides, the combinatorial and the probabilistic version of the inclusion–exclusion principle follow.

If one sees a number n as a set of its prime factors, then (**) is a generalization of Möbius inversion formula for square-free natural numbers. Therefore, (**) is seen as the Möbius inversion formula for the incidence algebra of the partially ordered set of all subsets of A.

For a generalization of the full version of Möbius inversion formula, (**) must be generalized to multisets. For multisets instead of sets, (**) becomes

$$f(A) = \sum_{S \subseteq A} \mu(A - S) g(S) \qquad (***)$$

where $A - S$ is the multiset for which $(A - S) \uplus S = A$, and

- $\mu(S) = 1$ if S is a set (i.e. a multiset without double elements) of even cardinality.
- $\mu(S) = -1$ if S is a set (i.e. a multiset without double elements) of odd cardinality.
- $\mu(S) = 0$ if S is a proper multiset (i.e. S has double elements).

Notice that $\mu(A - S)$ is just the $(-1)^{|A|-|S|}$ of (**) in case $A - S$ is a set.

Proof of (***)

Substitute

$$g(S) = \sum_{T \subseteq S} f(T)$$

on the right hand side of (***). Notice that $f(A)$ appears once on both sides of (***). So we must show that for all T with $T \subsetneq A$, the terms $f(T)$ cancel out on the right hand side of (***). For that purpose, take a fixed T such that $T \subsetneq A$ and take an arbitrary fixed $a \in A$ such that $a \notin T$.

Notice that $A - S$ must be a set for each positive or negative appearance of $f(T)$ on the right hand side of (***) that is obtained by way of the multiset S such that $T \subseteq S \subseteq A$. Now each appearance of $f(T)$ on the right hand side of (***) that is obtained by way of S such that $A - S$ is a set that contains a cancels out with the one that is obtained by way of the corresponding S such that $A - S$ is a set that does not contain a. This gives the desired result.

Applications

The inclusion–exclusion principle is widely used and only a few of its applications can be mentioned here.

Counting Derangements

A well-known application of the inclusion–exclusion principle is to the combinatorial problem of counting all derangements of a finite set. A *derangement* of a set A is a bijection from A into itself that has no fixed points. Via the inclusion–exclusion principle one can show that if the cardinality of A is n, then the number of derangements is $[n! / e]$ where $[x]$ denotes the nearest integer to x.

The first occurrence of the problem of counting the number of derangements is in an early book on games of chance: *Essai d'analyse sur les jeux de hazard* by P. R. de Montmort (1678 – 1719) and was known as either "Montmort's problem" or by the name he gave it, "*problème des rencontres.*" The problem is also known as the *hatcheck problem.*

The number of derangements is also known as the subfactorial of n, written $!n$. It follows that if all bijections are assigned the same probability then the probability that a random bijection is a derangement quickly approaches $1/e$ as n grows.

Counting Intersections

The principle of inclusion–exclusion, combined with De Morgan's law, can be used to count the cardinality of the intersection of sets as well. Let $\overline{A_k}$ represent the complement of A_k with respect to some universal set A such that $A_k \subseteq A$ for each k. Then we have

$$\bigcap_{i=1}^{n} A_i = \overline{\bigcup_{i=1}^{n} \overline{A_i}}$$

thereby turning the problem of finding an intersection into the problem of finding a union.

Graph Coloring

The inclusion exclusion principle forms the basis of algorithms for a number of NP-hard graph partitioning problems, such as graph coloring.

A well known application of the principle is the construction of the chromatic polynomial of a graph.

Bipartite Graph Perfect Matchings

The number of perfect matchings of a bipartite graph can be calculated using the principle.

Number of Onto Functions

Given finite sets A and B, how many surjective functions (onto functions) are there from A to B? Without any loss of generality we may take $A = \{1,2,\ldots,k\}$ and $B = \{1,2,\ldots,n\}$, since only the cardinalities of the sets matter. By using S as the set of all functions from A to B, and defining, for each i in B, the property P_i as "the function misses the element i in B" (i is not in the image of the function), the principle of inclusion–exclusion gives the number of onto functions between A and B as:

$$\sum_{j=0}^{n}\binom{n}{j}(-1)^j(n-j)^k.$$

Permutations with Forbidden Positions

A permutation of the set $S = \{1,2,...,n\}$ where each element of S is restricted to not being in certain positions (here the permutation is considered as an ordering of the elements of S) is called a *permutation with forbidden positions*. For example, with $S = \{1,2,3,4\}$, the permutations with the restriction that the element 1 can not be in positions 1 or 3, and the element 2 can not be in position 4 are: 2134, 2143, 3124, 4123, 2341, 2431, 3241, 3421, 4231 and 4321. By letting A_i be the set of positions that the element i is not allowed to be in, and the property P_i to be the property that a permutation puts element i into a position in A_i, the principle of inclusion–exclusion can be used to count the number of permutations which satisfy all the restrictions.

In the given example, there are 12 = 2(3!) permutations with property P_1, 6 = 3! permutations with property P_2 and no permutations have properties P_3 or P_4 as there are no restrictions for these two elements. The number of permutations satisfying the restrictions is thus:

$$4! - (12 + 6 + 0 + 0) + (4) = 24 - 18 + 4 = 10.$$

The final 4 in this computation is the number of permutations having both properties P_1 and P_2. There are no other non-zero contributions to the formula.

Stirling Numbers of the Second Kind

The Stirling numbers of the second kind, $S(n,k)$ count the number of partitions of a set of n elements into k non-empty subsets (indistinguishable *boxes*). An explicit formula for them can be obtained by applying the principle of inclusion–exclusion to a very closely related problem, namely, counting the number of partitions of an n-set into k non-empty but distinguishable boxes (ordered non-empty subsets). Using the universal set consisting of all partitions of the n-set into k (possibly empty) distinguishable boxes, $A_1, A_2, ..., A_k$, and the properties P_i meaning that the partition has box A_i empty, the principle of inclusion–exclusion gives an answer for the related result. Dividing by $k!$ to remove the artificial ordering gives the Stirling number of the second kind:

$$S(n,k) = \frac{1}{k!}\sum_{t=0}^{k}(-1)^t\binom{k}{t}(k-t)^n.$$

Rook Polynomials

A rook polynomial is the generating function of the number of ways to place non-attacking rooks on a *board B* that looks like a subset of the squares of a checkerboard; that is, no two rooks may be in the same row or column. The board B is any subset of the squares of a rectangular board with n rows and m columns; we think of it as the squares in which one is allowed to put a rook. The coefficient, $r_k(B)$ of x^k in the rook polynomial $R_B(x)$ is the number of ways k rooks, none of which attacks another, can be arranged in the squares of B. For any board B, there is a complementary board B' consisting of the squares of the rectangular board that are not in B. This complementary board also has a rook polynomial $R_{B'}(x)$ with coefficients $r_k(B')$.

It is sometimes convenient to be able to calculate the highest coefficient of a rook polynomial in terms of the coefficients of the rook polynomial of the complementary board. Without loss of generality we can assume that $n \leq m$, so this coefficient is $r_n(B)$. The number of ways to place n non-attacking rooks on the complete $n \times m$ "checkerboard" (without regard as to whether the rooks are placed in the squares of the board B) is given by the falling factorial:

$$(m)_n = m(m-1)(m-2)\cdots(m-n+1).$$

Letting P_i be the property that an assignment of n non-attacking rooks on the complete board has a rook in column i which is not in a square of the board B, then by the principle of inclusion–exclusion we have:

$$r_n(B) = \sum_{t=0}^{n} (-1)^t (m-t)_{n-t}\, r_t(B').$$

Euler's Phi Function

Euler's totient or phi function, $\varphi(n)$ is an arithmetic function that counts the number of positive integers less than or equal to n that are relatively prime to n. That is, if n is a positive integer, then $\varphi(n)$ is the number of integers k in the range $1 \leq k \leq n$ which have no common factor with n other than 1. The principle of inclusion–exclusion is used to obtain a formula for $\varphi(n)$. Let S be the set $\{1,2,...,n\}$ and define the property P_i to be that a number in S is divisible by the prime number p_i, for $1 \leq i \leq r$, where the prime factorization of,

$$n = p_1^{a_1} p_2^{a_2} \cdots p_r^{a_r}.$$

Then,

$$\varphi(n) = n - \sum_{i=1}^{r} \frac{n}{p_i} + \sum_{1 \leq i < j \leq r} \frac{n}{p_i p_j} - \cdots = n \prod_{i=1}^{r} \left(1 - \frac{1}{p_i}\right).$$

Diluted Inclusion–exclusion Principle

In many cases where the principle could give an exact formula (in particular, counting prime numbers using the sieve of Eratosthenes), the formula arising doesn't offer useful content because the number of terms in it is excessive. If each term individually can be estimated accurately, the accumulation of errors may imply that the inclusion–exclusion formula isn't directly applicable. In number theory, this difficulty was addressed by Viggo Brun. After a slow start, his ideas were taken up by others, and a large variety of sieve methods developed. These for example may try to find upper bounds for the "sieved" sets, rather than an exact formula.

Let $A_1, ..., A_n$ be arbitrary sets and $p_1, ..., p_n$ real numbers in the closed unit interval $[0,1]$. Then, for every even number k in $\{0, ..., n\}$, the indicator functions satisfy the inequality:

$$\mathbf{1}_{A_1\cup\cdots\cup A_n} \geq \sum_{j=1}^{k}(-1)^{j-1}\sum_{1\leq i_1<\cdots<i_j\leq n} p_{i_1}\cdots p_{i_j}\mathbf{1}_{A_{i_1}\cap\cdots\cap A_{i_j}}.$$

Proof of Main Statement

Choose an element contained in the union of all sets and let $A_1, A_2, \ldots, A_t$ be the individual sets containing it. (Note that $t > 0$.) Since the element is counted precisely once by the left-hand side of earlier equation, we need to show that it is counted precisely once by the right-hand side. On the right-hand side, the only non-zero contributions occur when all the subsets in a particular term contain the chosen element, that is, all the subsets are selected from $A_1, A_2, \ldots, A_t$. The contribution is one for each of these sets (plus or minus depending on the term) and therefore is just the (signed) number of these subsets used in the term. We then have:

$$\begin{aligned}&|\{A_i \mid 1\leq i\leq t\}| - |\{A_i\cap A_j \mid 1\leq i<j\leq t\}| + \cdots \\ &\cdots + (-1)^{t+1}|\{A_1\cap A_2\cap\cdots\cap A_t\}| \\ &= \binom{t}{1}-\binom{t}{2}+\cdots+(-1)^{t+1}\binom{t}{t}.\end{aligned}$$

By the binomial theorem,

$$0=(1-1)^t=\binom{t}{0}-\binom{t}{1}+\binom{t}{2}-\cdots+(-1)^t\binom{t}{t}.$$

Using the fact that $\binom{t}{0}=1$ and rearranging terms, we have

$$1=\binom{t}{1}-\binom{t}{2}+\cdots+(-1)^{t+1}\binom{t}{t},$$

and so, the chosen element is counted only once by the earlier equation.

Algebraic Proof

An algebraic proof can be obtained using indicator functions (characteristic functions of subsets of a set). The indicator function of a subset S of a set X is a function,

$$\mathbf{1}_S : X \to \{0,1\}$$

defined as,

$$\mathbf{1}_S(x) := \begin{cases}1 & \text{if } x\in S,\\ 0 & \text{if } x\notin S.\end{cases}$$

If A and B are two subsets of X, then

$$\mathbf{1}_A \cdot \mathbf{1}_B = \mathbf{1}_{A \cap B}.$$

Let A denote the union $\bigcup_{i=1}^{n} A_i$ of the sets $A_1, ..., A_n$. To prove the inclusion–exclusion principle in general, we first verify the identity

$$1_A = \sum_{k=1}^{n} (-1)^{k-1} \sum_{\substack{I \subset \{1,...,n\} \\ |I|=k}} 1_{A_I} \qquad (*)$$

for indicator functions, where

$$A_I = \bigcap_{i \in I} A_i.$$

The following function is identically zero,

$$(1_A - 1_{A_1})(1_A - 1_{A_2}) \cdots (1_A - 1_{A_n}) = 0,$$

because: if x is not in A, then all factors are 0 – 0 = 0; and otherwise, if x does belong to some A_m, then the corresponding mth factor is 1 – 1 = 0. By expanding the product on the left-hand side, equation (∗) follows.

To prove the inclusion–exclusion principle for the cardinality of sets, sum the equation (∗) over all x in the union of $A_1, ..., A_n$. To derive the version used in probability, take the expectation in (∗). In general, integrate the equation (∗) with respect to μ. Always use linearity in these derivations.

Principle of Inclusion and Exclusion

The following result is well known and hence we omit the proof.

Theorem: Let U be a finite set. Suppose A and B are two subsets of U. Then the number of elements of U that are neither in A nor in B are,

$$|U| - (|A| + |B| - |A \cap B|).$$

Or equivalently, $|A \cup B| = |A| + |B| - |A \cap B|$.

A generalization of this to three subsets A, B and C is also well known. To get a result that generalizes Theorem for n subsets $A_1, A_2, \ldots, A_n$, we need the following notations:

$$S_1 = \sum_{i=1}^{n} |A_i|, S_2 = \sum_{1 \le i < j \le n} |A_i \cap A_j|, S_3 = \sum_{1 \le i < j < k \le n} |A_i \cap A_j \cap A_k|, \cdots, S_n = |A_1 \cap A_2 \cap \cdots \cap A_n|.$$

With the notations as defined above, we have the following theorem, called the inclusion- exclusion principle. This theorem can be easily proven using the principle of mathematical induction. But we give a separate proof for better understanding.

Theorem: [Inclusion-Exclusion Principle] Let $A_1, A_2, \ldots, A_n$ be n subsets of a finite set U. Then the number of elements of U that are in none of $A_1, A_2, \ldots, A_n$ is given by,

$$|U| - S_1 + S_2 - S_3 + \cdots + (-1)^n S_n. \qquad (1)$$

Or equivalently,

$$|A_1 \cup A_2 \cup \cdots \cup A_n| = S_1 - S_2 + \cdots + (-1 \quad S_n. \qquad (2)$$

Proof. We show that if an element $x \in U$ belongs to exactly k of the subsets $A_1, A_2, \ldots, A_n$, for some $k \geq 1$, then its contribution in (1) is zero. Suppose x belongs to exactly k subsets $A_{i_1}, A_{i_2}, \ldots, A_{i_k}$. Then we observe the following:

1. The contribution of x *in* $|U|$ *is* 1.
2. The contribution of x *in* S_1 *is* k *as* $x \in A_{i_j}, 1 \leq j \leq k$.
3. The contribution of x *in* S_2 *is* $\binom{k}{2}$ *as* $x \in A_{i_j} \cap A_{i_l}, 1 \leq j < l \leq k$.
4. The contribution of x *in* S_3 *is* $\binom{k}{3}$ *as* $x \in A_{i_j} \cap A_{i_l} \cap A_{i_m}, 1 \leq j < l < m \leq k$.

Proceeding this way, we have

5. The contribution of x *in* S_k *is* $\binom{k}{k} = 1$, and
6. The contribution of x *in* S_ℓ *for* $\ell \geq k+1$ is 0.

So, the contribution of x in (1) is,

$$1 - k + \binom{k}{2} - \binom{k}{3} + \cdots + (-1)^{k-1}\binom{k}{k-1} + (-1)^k \binom{k}{k} + 0 \cdots + 0 = (1-1)^k = 0.$$

This completes the proof of the theorem. The readers are advised to prove the equivalent condition.

Example: 1. Determine the number of 10-letter words using ENGLISH alphabets that does not contain all the vowels.

Solution: Let U be the set consisting of all the 10-letters words using ENGLISH alphabets and let Aα be a subset of U that does not contain the letter α. Then we need to compute,

$$|A_a \cup A_e \cup A_i \cup A_o \cup A_u| = S_1 - S_2 + S_3 - S_4 + S_5,$$

where $S_1 = \sum_{\alpha \in \{a,e,i,o,u\}} |A_\alpha| = \binom{5}{1} 25^{10}, S_2 = \binom{5}{2} 24^{10}, S_3 = \binom{5}{3} 23^{10}, S_4 = \binom{5}{4} 22^{10}$ *and* $S_5 = 21^{10}$. So, the required answer is $\sum_{k=1}^{5} (-1)^{k-1} \binom{5}{k} (26-k)^{10}$.

2. Determine the number of integers between 1 and 1000 that are coprime to 2, 3, 11 and 13. Solution: Let U = {1, 2, 3, . . . , 1000} and let A_i = {n ∈ U : i divides n}, for i = 2, 3, 11, 13. Then note that we need the value of $|U| - |A_2 \cup A_3 \cup A_{11} \cup A_{13}|$. Observe that

$$|A_2| = \left\lfloor \frac{1000}{2} \right\rfloor = 500, |A_3| = \left\lfloor \frac{1000}{3} \right\rfloor = 333, |A_{11}| = \left\lfloor \frac{1000}{11} \right\rfloor = 90, |A_{13}| = \left\lfloor \frac{1000}{13} \right\rfloor = 76,$$

$$|A_2 \cap A_3| = \left\lfloor \frac{1000}{6} \right\rfloor = 166, |A_2 \cap A_{11}| = \left\lfloor \frac{1000}{22} \right\rfloor = 45, |A_2 \cap A_{13}| = \left\lfloor \frac{1000}{26} \right\rfloor = 38,$$

$$|A_3 \cap A_{11}| = \left\lfloor \frac{1000}{33} \right\rfloor = 30, |A_3 \cap A_{13}| = \left\lfloor \frac{1000}{39} \right\rfloor = 25, |A_{11} \cap A_{13}| = \left\lfloor \frac{1000}{143} \right\rfloor = 6,$$

$$|A_2 \cap A_3 \cap A_{11}| = 15, |A_2 \cap A_3 \cap A_{13}| = 12, |A_2 \cap A_{11} \cap A_{13}| = 3,$$

$$|A_3 \cap A_{11} \cap A_{13}| = 2, |A_2 \cap A_3 \cap A_{11} \cap A_{13}| = 1.$$

Thus, the required number is,

1000−((500+333+90+76)−(166+45+38+30+25+6)−(15+12+3+2)−1) = 1000−720 = 280.

3. (Euler's ϕ-function Or Euler's totient function) Let n denote a positive integer.Then the Euler ϕ-function is defined by,

$$\phi(n) = \left|\{k\ : 1 \le k \le n,\ \gcd(n,k) = 1\}\right|. \qquad (3)$$

Determine a formula for $\phi(n)$ in terms of its prime factors.

Solution: Let $n = p_1^{\alpha 1} p_2^{\alpha 2} \cdots p_k^{\alpha k}$ be the unique decomposition of n as product of distinct primes $p_1, p_2, \ldots, p_k$, $U = \{1, 2, \ldots, n\}$ and let $A_{p_i} = \{m \in U : p_i \text{ divides } m\}, \text{ for } 1 \le i \le k$ Then, by definition,

$$\begin{aligned}\phi(n) &= |U| - S_1 + S_2 - S_3 + \cdots + (-1)^k S_k \\ &= n - \sum_{i=1}^{k} \frac{n}{p_i} + \sum_{1 \le i < j \le k} \frac{n}{p_i p_j} - \cdots + (-1)^k \frac{n}{p_1 p_2 \cdots p_k} \\ &= n \prod_{i=1}^{k} \left(1 - \frac{1}{p_i}\right) \end{aligned} \qquad (4)$$

Formal Power Series

In mathematics, a formal power series is a generalization of a polynomial, where the number of terms is allowed to be infinite; this implies giving up the possibility of replacing the variable in the polynomial with an arbitrary number. Thus a formal power series differs from a polynomial in that it may have infinitely many terms, and differs from a power series, whose variables can take on numerical values. One way to view a formal power series is as an infinite ordered sequence of numbers. In this case, the powers of the variable are used only to indicate the order of the coefficients, so that the coefficient of x^5 is the fifth term in the sequence. In combinatorics, formal power series provide representations of numerical sequences and of multisets, and for instance allow concise expressions for recursively defined sequences regardless of whether the recursion can be explicitly solved; this is known as the method of generating functions. More generally, formal power series can include series with any finite number of variables, and with coefficients in an arbitrary ring.

Introduction

A formal power series can be loosely thought of as an object that is like a polynomial, but with infinitely many terms. Alternatively, for those familiar with power series (or Taylor series), one may think of a formal power series as a power series in which we ignore questions of convergence by not assuming that the variable X denotes any numerical value (not even an unknown value). For example, consider the series,

$$A = 1 - 3X + 5X^2 - 7X^3 + 9X^4 - 11X^5 + \cdots.$$

If we studied this as a power series, its properties would include, for example, that its radius of convergence is 1. However, as a formal power series, we may ignore this completely; all that is relevant is the sequence of coefficients [1, −3, 5, −7, 9, −11, ...]. In other words, a formal power series is an object that just records a sequence of coefficients. It is perfectly acceptable to consider a formal power series with the factorials [1, 1, 2, 6, 24, 120, 720, 5040, ...] as coefficients, even though the corresponding power series diverges for any nonzero value of X.

Arithmetic on formal power series is carried out by simply pretending that the series are polynomials. For example, if

$$B = 2X + 4X^3 + 6X^5 + \cdots,$$

then we add A and B term by term:

$$A + B = 1 - X + 5X^2 - 3X^3 + 9X^4 - 5X^5 + \cdots.$$

We can multiply formal power series, again just by treating them as polynomials,

$$AB = 2X - 6X^2 + 14X^3 - 26X^4 + 44X^5 + \cdots.$$

Notice that each coefficient in the product AB only depends on a *finite* number of coefficients of A and B. For example, the X^5 term is given by,

$$44X^5 = (1 \times 6X^5) + (5X^2 \times 4X^3) + (9X^4 \times 2X).$$

For this reason, one may multiply formal power series without worrying about the usual questions of absolute, conditional and uniform convergence which arise in dealing with power series in the setting of analysis.

Once we have defined multiplication for formal power series, we can define multiplicative inverses as follows. The multiplicative inverse of a formal power series A is a formal power series C such that $AC = 1$, provided that such a formal power series exists. It turns out that if A has a multiplicative inverse, it is unique, and we denote it by A^{-1}. Now we can define division of formal power series by defining B/A to be the product BA^{-1}, provided that the inverse of A exists. For example, one can use the definition of multiplication above to verify the familiar formula,

$$\frac{1}{1+X} = 1 - X + X^2 - X^3 + X^4 - X^5 + \cdots.$$

An important operation on formal power series is coefficient extraction. In its most basic form, the coefficient extraction operator for a formal power series in one variable extracts the coefficient of $[X^n]$, and is written e.g. $[X^n]A$, so that $[X^2]A = 5$ and $[X^5]A = -11$. Other examples include,

$$\begin{aligned}
\left[X^3\right](B) &= 4, \\
\left[X^2\right](X + 3X^2Y^3 + 10Y^6) &= 3Y^3, \\
\left[X^2Y^3\right](X + 3X^2Y^3 + 10Y^6) &= 3, \\
\left[X^n\right]\left(\frac{1}{1+X}\right) &= (-1)^n, \\
\left[X^n\right]\left(\frac{X}{(1-X)^2}\right) &= n.
\end{aligned}$$

Similarly, many other operations that are carried out on polynomials can be extended to the formal power series setting, as explained below.

The Ring of Formal Power Series

The set of all formal power series in X with coefficients in a commutative ring R form another ring that is written $R[[X]]$, and called the ring of formal power series in the variable X over R.

Definition of the Formal Power Series Ring

One can characterize $R[[X]]$ abstractly as the completion of the polynomial ring $R[X]$ equipped with a particular metric. This automatically gives $R[[X]]$ the structure of a topological ring (and even of a complete metric space). But the general construction of a completion of a metric space is more involved than what is needed here, and would make formal power series seem more complicated than they are. It is possible to describe $R[[X]]$ more explicitly, and define the ring structure and topological structure separately, as follows.

Ring Structure

As a set, $R[[X]]$ can be constructed as the set $R^{\mathbb{N}}$ of all infinite sequences of elements of R, indexed by the natural numbers (taken to include 0). Designating a sequence whose term at index n is a_n by (a_n), one defines addition of two such sequences by,

$$(a_n)_{n\in\mathbb{N}} + (b_n)_{n\in\mathbb{N}} = (a_n + b_n)_{n\in\mathbb{N}}$$

and multiplication by,

$$(a_n)_{n\in\mathbb{N}} \times (b_n)_{n\in\mathbb{N}} = \left(\sum_{k=0}^{n} a_k b_{n-k}\right)_{n\in\mathbb{N}}.$$

This type of product is called the Cauchy product of the two sequences of coefficients, and is a sort of discrete convolution. With these operations, $R^{\mathbb{N}}$ becomes a commutative ring with zero element $(0,0,0,\cdots)$ and multiplicative identity $(1,0,0,\cdots)$.

The product is in fact the same one used to define the product of polynomials in one indeterminate, which suggests using a similar notation. One embeds R into $R[[X]]$ by sending any (constant) $a \in R$ to the sequence $(a,0,0,\cdots)$ and designates the sequence $(0,1,0,0,\cdots)$ by X; then using the above definitions every sequence with only finitely many nonzero terms can be expressed in terms of these special elements as,

$$(a_0, a_1, a_2, \cdots, a_n, 0, 0, \ldots) = a_0 + a_1 X + \cdots + a_n X^n = \sum_{i=0}^{n} a_i X^i;$$

these are precisely the polynomials in X. Given this, it is quite natural and convenient to designate a general sequence $(a_n)_{n\in\mathbb{N}}$ by the formal expression $\sum_{i\in\mathbb{N}} a_i X^i$, even though the latter *is not* an expression formed by the operations of addition and multiplication defined above (from which only finite sums can be constructed). This notational convention allows reformulation the above definitions as,

$$\left(\sum_{i\in\mathbb{N}} a_i X^i\right) + \left(\sum_{i\in\mathbb{N}} b_i X^i\right) = \sum_{i\in\mathbb{N}} (a_i + b_i) X^i$$

and

$$\left(\sum_{i\in\mathbb{N}} a_i X^i\right) \times \left(\sum_{i\in\mathbb{N}} b_i X^i\right) = \sum_{n\in\mathbb{N}} \left(\sum_{k=0}^{n} a_k b_{n-k}\right) X^n.$$

which is quite convenient, but one must be aware of the distinction between formal summation (a mere convention) and actual addition.

Topological Structure

Having stipulated conventionally that,

$$(a_0, a_1, a_2, a_3, \ldots) = \sum_{i=0}^{\infty} a_i X^i,$$

one would like to interpret the right hand side as a well-defined infinite summation. To that end, a notion of convergence in $R^{\mathbb{N}}$ is defined and a topology on $R^{\mathbb{N}}$ is constructed. There are several equivalent ways to define the desired topology.

- We may give $R^{\mathbb{N}}$ the product topology, where each copy of R is given the discrete topology.
- We may give $R^{\mathbb{N}}$ the I-adic topology, where $I=(X)$ is the ideal generated by X, which consists of all sequences whose first term a_0 is zero.
- The desired topology could also be derived from the following metric. The distance between distinct sequences (a_n) and (b_n) in $R^{\mathbb{N}}$, is defined to be.

$$d((a_n),(b_n)) = 2^{-k},$$

where k is the smallest natural number such that $a_k \neq b_k$; the distance between two equal sequences is of course zero.

Informally, two sequences $\{a_n\}$ and $\{b_n\}$ become closer and closer if and only if more and more of their terms agree exactly. Formally, the sequence of partial sums of some infinite summation converges if for every fixed power of X the coefficient stabilizes: there is a point beyond which all further partial sums have the same coefficient. This is clearly the case for the right hand side, regardless of the values a_n, since inclusion of the term for $\{i=n\}$ gives the last (and in fact only) change to the coefficient of X^n. It is also obvious that the limit of the sequence of partial sums is equal to the left hand side.

This topological structure, together with the ring operations described above, form a topological ring. This is called the ring of formal power series over R and is denoted by $R[[X]]$. The topology has the useful property that an infinite summation converges if and only if the sequence of its terms converges to 0, which just means that any fixed power of X occurs in only finitely many terms.

The topological structure allows much more flexible use of infinite summations. For instance the rule for multiplication can be restated simply as.

$$\left(\sum_{i\in\mathbb{N}} a_i X^i\right)\times\left(\sum_{i\in\mathbb{N}} b_i X^i\right)=\sum_{i,j\in\mathbb{N}} a_i b_j X^{i+j},$$

since only finitely many terms on the right affect any fixed X^n. Infinite products are also defined by the topological structure; it can be seen that an infinite product converges if and only if the sequence of its factors converges to 1.

Alternative Topologies

The above topology is the finest topology for which $\sum_{i=0}^{\infty} a_i X^i$ always converges as a summation to the formal power series designated by the same expression, and it often suffices to give a meaning to infinite sums and products, or other kinds of limits that one wishes to use to designate particular formal power series. It can however happen occasionally that one wishes to use a coarser topology, so that certain expressions become convergent that would otherwise diverge. This applies in particular when the base ring $\mathbb{R}$ already comes with a topology other than the discrete one, for instance if it is also a ring of formal power series.

Consider the ring of formal power series,

$$\mathbb{Z}[[X]][[Y]]$$

then the topology of above construction only relates to the indeterminate Y, since the topology that was put on $\mathbb{Z}[[X]]$ has been replaced by the discrete topology when defining the topology of the whole ring. So

$$\sum_{i \in \mathbb{N}} XY^i$$

converges to the power series suggested, which can be written as $\frac{X}{1-Y}$; however the summation,

$$\sum_{i \in \mathbb{N}} X^i Y$$

would be considered to be divergent, since every term affects the coefficient of Y (which coefficient is itself a power series in X). This asymmetry disappears if the power series ring in Y is given the product topology where each copy of $\mathbb{Z}[[X]]$ is given its topology as a ring of formal power series rather than the discrete topology. As a consequence, for convergence of a sequence of elements of $\mathbb{Z}[[X]][[Y]]$ it then suffices that the coefficient of each power of Y converges to a formal power series in X, a weaker condition that stabilizing entirely; for instance in the second example given here the coefficient of Y converges to $\frac{1}{1-X}$, so the whole summation converges to $\frac{Y}{1-X}$.

This way of defining the topology is in fact the standard one for repeated constructions of rings of formal power series, and gives the same topology as one would get by taking formal power series in all inderteminates at once. In the above example that would mean constructing $\mathbb{Z}[[X,Y]]$, and here a sequence converges if and only if the coefficient of every monomial X^iY^j stabilizes. This topology, which is also the I-adic topology, where $I=(X,Y)$ is the ideal generated by X and Y, still enjoys the property that a summation converges if and only if its terms tend to 0.

The same principle could be used to make other divergent limits converge. For instance in $\mathbb{R}[[X]]$ the limit,

$$\lim_{n\to\infty}\left(1+\frac{X}{n}\right)^n$$

does not exist, so in particular it does not converge to $\exp(X)=\sum_{n\in\mathbb{N}}\frac{X^n}{n!}$. This is because for $i \geq 2$ the coefficient $\binom{n}{i}/n^i$ of X^i does not stabilize as $n \to \infty$. It does however converge in the usual topology of $\mathbb{R}$, and in fact to the coefficient $\frac{1}{i!}$ of $\exp(X)$. Therefore, if one would give $\mathbb{R}[[X]]$ the product topology of $\mathbb{R}^{\mathbb{N}}$ where the topology of $\mathbb{R}$ is the usual topology rather than the discrete one, then the above limit would converge to $\exp(X)$. This more permissive approach is not however the standard when considering formal power series, as it would lead to convergence considerations that are as subtle as they are in analysis, while the philosophy of formal power series is on the contrary to make convergence questions as trivial as they can possibly be. With this topology it would *not* be the case that a summation converges if and only if its terms tend to 0.

Universal Property

The ring $\mathbb{R}[[X]]$ may be characterized by the following universal property. If S is a commutative associative algebra over R, if I is an ideal of S such that the I-adic topology on S is complete, and if x is an element of I, then there is a *unique* $\Phi : R[[X]] \to S$ with the following properties:

- Φ is an R-algebra homomorphism
- Φ is continuous
- $\Phi(X) = x$

Operations on Formal Power Series

One can perform algebraic operations on power series to generate new power series. Besides the ring structure operations defined above, we have the following:

Power Series Raised to Powers

If n is a natural number we have,

$$\left(\sum_{k=0}^{\infty} a_k X^k\right)^n = \sum_{m=0}^{\infty} c_m X^m,$$

where

$$c_0 = a_0^n, \quad c_m = \frac{1}{m a_0} \sum_{k=1}^{m} (kn - m + k) a_k c_{m-k},$$

for $m \geq 1$. (This formula can only be used if m and a_0 are invertible in the ring of scalars.)

In the case of formal power series with complex coefficients, the complex powers are well defined at least for series f with constant term equal to 1. In this case, f^α can be defined either by composition with the binomial series $(1+x)^\alpha$, or by composition with the exponential and the logarithmic series, $f^\alpha := \exp(\alpha \log(f))$, or as the solution of the differential equation $f(f^\alpha)' = \alpha f^\alpha f'$ with constant term 1, the three definitions being equivalent. The rules of calculus $(f^\alpha)^\beta = f^{\alpha\beta}$ and $f^\alpha g^\alpha = (fg)^\alpha$ easily follow.

Inverting Series

The series

$$A = \sum_{n=0}^{\infty} a_n X^n$$

in $\mathbb{R}[[X]]$ is invertible in $\mathbb{R}[[X]]$ if and only if its constant coefficient a_0 is invertible in R. This condition is necessary, for the following reason: if we suppose that A has an inverse $B = b_0 + b_1 x + \cdots$

then the constant term $a_0 b_0$ of $A \cdot B$ is the constant term of the identity series, i.e., it is 1. This condition is also sufficient; we may compute the coefficients of the inverse series B via the explicit recursive formula,

$$b_0 = \frac{1}{a_0}$$

$$b_n = -\frac{1}{a_0} \sum_{i=1}^{n} a_i b_{n-i} \qquad \text{for } n \geq 1.$$

An important special case is that the geometric series formula is valid in $K[[X]]$:

$$(1-X)^{-1} = \sum_{n=0}^{\infty} X^n.$$

If $R = K$ is a field, then a series is invertible if and only if the constant term is non-zero, i.e., if and only if the series is not divisible by X. This says that $K[[X]]$ is a discrete valuation ring with uniformizing parameter X.

Dividing Series

The computation of a quotient $f/g = h$

$$\frac{\sum_{n=0}^{\infty} b_n X^n}{\sum_{n=0}^{\infty} a_n X^n} = \sum_{n=0}^{\infty} c_n X^n,$$

assuming the denominator is invertible (that is, a_0 is invertible in the ring of scalars), can be performed as a product f and the inverse of g, or directly equating the coefficients in $f = gh$:

$$c_n = \frac{1}{a_0}\left(b_n - \sum_{k=1}^{n} a_k c_{n-k}\right).$$

Extracting Coefficients

The coefficient extraction operator applied to a formal power series,

$$f(X) = \sum_{n=0}^{\infty} a_n X^n$$

in X is written

$$\left[X^m\right] f(X)$$

and extracts the coefficient of X^m, so that

$$[X^m]f(X)=[X^m]\sum_{n=0}^{\infty}a_nX^n=a_m.$$

Composition of Series

Given formal power series,

$$f(X)=\sum_{n=1}^{\infty}a_nX^n=a_1X+a_2X^2+\cdots$$

$$g(X)=\sum_{n=0}^{\infty}b_nX^n=b_0+b_1X+b_2X^2+\cdots,$$

one may form the *composition*,

$$g(f(X))=\sum_{n=0}^{\infty}b_n(f(X))^n=\sum_{n=0}^{\infty}c_nX^n,$$

where the coefficients c_n are determined by "expanding out" the powers of $f(X)$:

$$c_n:=\sum_{k\in\mathbb{N},|j|=n}b_ka_{j_1}a_{j_2}\cdots a_{j_k}.$$

Here the sum is extended over all (k, j) with k in N and $j\in\mathbb{N}_+^k$ with $|j|:=j_1+\cdots+j_k=n$.

A more explicit description of these coefficients is provided by Faà di Bruno's formula, at least in the case where the coefficient ring is a field of characteristic 0.

A point here is that this operation is only valid when $f(X)$ has *no constant term*, so that each c_n depends on only a finite number of coefficients of $f(X)$ and $g(X)$. In other word the series for $g(f(X))$ converges in the topology of $R[[X]]$.

Example

Assume that the ring R has characteristic 0. If we denote by $\exp(X)$ the formal power series,

$$\exp(X)=1+X+\frac{X^2}{2!}+\frac{X^3}{3!}+\frac{X^4}{4!}+\cdots,$$

then the expression,

$$\exp(\exp(X)-1)=1+X+X^2+\frac{5X^3}{6}+\frac{5X^4}{8}+\cdots$$

makes perfect sense as a formal power series. However, the statement:

$$\exp(\exp(X))=e\exp(\exp(X)-1)=e+eX+eX^2+\frac{5eX^3}{6}+\cdots$$

is not a valid application of the composition operation for formal power series. Rather, it is confusing the notions of convergence in $R[[X]]$ and convergence in R; indeed, the ring R may not even contain any number e with the appropriate properties.

Composition Inverse

Whenever a formal series $f(X)=\sum_k f_k X^k \in R[[X]]$ has $f_0 = 0$ and f_1 being an invertible element of R, there exists a series $g(X)=\sum_k g_k X^k$ that is the composition inverse of f, meaning that composing f with g gives the series representing the identity function (whose first coefficient is 1 and all other coefficients are zero). The coefficients of g may be found recursively by using the above formula for the coefficients of a composition, equating them with those of the composition identity X (that is 1 at degree 1 and 0 at every degree greater than 1). In the case when the coefficient ring is a field of characteristic 0, the Lagrange inversion formula provides a powerful tool to compute the coefficients of g, as well as the coefficients of the (multiplicative) powers of g.

Formal Differentiation of Series

Given a formal power series,

$$f=\sum_{n\geq 0} a_n X^n$$

in $R[[X]]$, we define its formal derivative, denoted Df or f', by

$$Df=\sum_{n\geq 1} a_n n X^{n-1}.$$

The symbol D is called the formal differentiation operator. The motivation behind this definition is that it simply mimics term-by-term differentiation of a polynomial.

This operation is R-linear:

$$D(af+bg)=a\cdot Df+b\cdot Dg$$

for any a, b in R and any f, g in $R[[X]]$. Additionally, the formal derivative has many of the properties of the usual derivative of calculus. For example, the product rule is valid:

$$D(fg)=f\cdot(Dg)+(Df)\cdot g,$$

and the chain rule works as well:

$$D(f\circ g)=(Df\circ g)\cdot Dg,$$

whenever the appropriate compositions of series are defined.

Thus, in these respects formal power series behave like Taylor series. Indeed, for the f defined above, we find that,

$$(D^k f)(0) = k!a_k,$$

where D^k denotes the kth formal derivative (that is, the result of formally differentiating k times).

Properties

Algebraic Properties of the Formal Power Series Ring

$R[[X]]$ is an associative algebra over R which contains the ring $R[X]$ of polynomials over R; the polynomials correspond to the sequences which end in zeros.

The Jacobson radical of $R[[X]]$ is the ideal generated by X and the Jacobson radical of R; this is implied by the element invertibility criterion discussed above.

The maximal ideals of $R[[X]]$ all arise from those in R in the following manner: an ideal M of $R[[X]]$ is maximal if and only if $M \cap R$ is a maximal ideal of R and M is generated as an ideal by X and $M \cap R$.

Several algebraic properties of R are inherited by $R[[X]]$:

- if R is a local ring, then so is $R[[X]]$;
- if R is Noetherian, then so is $R[[X]]$; this is a version of the Hilbert basis theorem;
- if R is an integral domain, then so is $R[[X]]$;
- if K is a field, then $K[[X]]$ is a discrete valuation ring.

Topological Properties of the Formal Power Series Ring

The metric space ($R[[X]]$, d) is complete.

The ring $R[[X]]$ is compact if and only if R is finite. This follows from Tychonoff's theorem and the characterisation of the topology on $R[[X]]$ as a product topology.

Applications

Formal power series can be used to solve recurrences occurring in number theory and combinatorics. For an example involving finding a closed form expression for the Fibonacci numbers.

One can use formal power series to prove several relations familiar from analysis in a purely algebraic setting. Consider for instance the following elements of Q[[X]]:

$$\sin(X) := \sum_{n\geq 0} \frac{(-1)^n}{(2n+1)!} X^{2n+1}$$

$$\cos(X) := \sum_{n\geq 0} \frac{(-1)^n}{(2n)!} X^{2n}$$

Then one can show that,

$$\sin^2(X)+\cos^2(X)=1,$$
$$\frac{\partial}{\partial X}\sin(X)=\cos(X),$$
$$\sin(X+Y)=\sin(X)\cos(Y)+\cos(X)\sin(Y).$$

The last one being valid in the ring Q[[X,Y]].

For K a field, the ring $K[[X_1, ..., X_r]]$ is often used as the "standard, most general" complete local ring over K in algebra.

Interpreting Formal Power Series as Functions

In mathematical analysis, every convergent power series defines a function with values in the real or complex numbers. Formal power series can also be interpreted as functions, but one has to be careful with the domain and codomain. If $f = \sum a_n X^n$ is an element of $R[[X]]$, S is a commutative associative algebra over R, I is an ideal in S such that the I-adic topology on S is complete, and x is an element of I, then we can define

$$f(X)=\sum_{n\geq 0} a_n X^n.$$

This latter series is guaranteed to converge in S given the above assumptions on X. Furthermore, we have

$$(f+g)(X)=f(X)+g(X)$$

and

$$(fg)(X)=f(X)g(X).$$

Unlike in the case of bona fide functions, these formulas are not definitions but have to be proved.

Since the topology on $R[[X]]$ is the (X)-adic topology and $R[[X]]$ is complete, we can in particular apply power series to other power series, provided that the arguments don't have constant coefficients (so that they belong to the ideal (X)): $f(0)$, $f(X^2–X)$ and $f((1–X)^{-1} – 1)$ are all well defined for any formal power series $f\in R[[X]]$.

With this formalism, we can give an explicit formula for the multiplicative inverse of a power series f whose constant coefficient $a = f(0)$ is invertible in R:

$$f^{-1}=\sum_{n\geq 0} a^{-n-1}(a-f)^n.$$

If the formal power series g with $g(0) = 0$ is given implicitly by the equation,

$$f(g) = X$$

where f is a known power series with $f(0) = 0$, then the coefficients of g can be explicitly computed using the Lagrange inversion formula.

Generalizations

Formal Laurent Series

A formal Laurent series over a ring R is defined in a similar way to a formal power series, except that we also allow finitely many terms of negative degree (this is different from the classical Laurent series) that is series of the form,

$$f = \sum_{n \in \mathbb{Z}} a_n X^n$$

where $a_n = 0$ for all but finitely many negative indices n. Multiplication of such series can be defined. Indeed, similarly to the definition for formal power series, the coefficient of X^k of two series with respective sequences of coefficients $\{a_n\}$ and $\{b_n\}$ is,

$$\sum_{i \in \mathbb{Z}} a_i b_{k-i},$$

which sum is effectively finite because of the assumed vanishing of coefficients at sufficiently negative indices, and which sum zero for sufficiently negative k for the same reason.

For a non-zero formal Laurent series, the minimal integer n such that $a_n \neq 0$ is called the order of f, denoted ord (f). (The order of the zero series is $+\infty$.) The formal Laurent series form the ring of formal Laurent series over R, denoted by $R((X))$. It is equal to the localization of $R[[X]]$ with respect to the set of positive powers of X. It is a topological ring with the metric $d(f,g) = 2^{-\text{ord}(f-g)}$.

If $R = K$ is a field, then $K((X))$ is in fact a field, which may alternatively be obtained as the field of fractions of the integral domain $K[[X]]$.

One may define formal differentiation for formal Laurent series in a natural way (term-by-term). Precisely, the formal derivative of the formal Laurent series f above is,

$$f' = Df = \sum_{n \in \mathbb{Z}} n a_n X^{n-1}$$

which is again an element of $K((X))$. Notice that if f is a non-constant formal Laurent series, and K is a field of characteristic 0, then one has,

$$\text{ord}(f') = \text{ord}(f) - 1.$$

However, in general this is not the case since the factor n for the lowest order term could be equal to 0 in R.

Formal Residue

Assume that K is a field of characteristic 0. Then the map,

$$D: K((X)) \to K((X))$$

is a K-derivation that satisfies,

$$\ker D = K$$
$$\operatorname{im} D = \{f \in K((X)) : [X^{-1}]f = 0\}.$$

The latter shows that the coefficient of X^{-1} in f is of particular interest; it is called *formal residue of f* and denoted $\operatorname{Res}(f)$. The map

$$\operatorname{Res}: K((X)) \to K$$

is K-linear, and by the above observation one has an exact sequence,

$$0 \to K \to K((X)) \xrightarrow{D} K((X)) \xrightarrow{\operatorname{Res}} K \to 0.$$

Some rules of calculus. As a quite direct consequence of the above definition, and of the rules of formal derivation, one has, for any $f, g \in K((X))$

i. $\operatorname{Res}(f') = 0;$

ii. $\operatorname{Res}(fg') = -\operatorname{Res}(f'g);$

iii. $\operatorname{Res}(f'/f) = \operatorname{ord}(f), \quad \forall f \neq 0;$

iv. $\operatorname{Res}((f \circ g)g') = \operatorname{ord}(g)\operatorname{Res}(f)$, if $\operatorname{ord}(g) > 0;$

v. $[X^n]f(X) = \operatorname{Res}\left(X^{-n-1}f(X)\right).$

Property (i) is part of the exact sequence above. Property (ii) follows from (i) as applied to $(fg)' = f'g + fg'$. Property (iii): any f can be written in the form $f = x^m g$, with $m = \operatorname{ord}(f)$ and $\operatorname{ord}(g) = 0$: then $f'/f = mX^{-1} + g'/g$. Since $\operatorname{ord}(g) = 0$, the element g is invertible in $K[[X]] \subset \operatorname{im}(D) = \ker(\operatorname{Res})$, whence $\operatorname{Res}(f'/f) = m$. Property (iv): Since $\operatorname{im}(D) = \ker(\operatorname{Res})$, we can write $f = f_{-1}X^{-1} + F'$, with $F \in K[[X]]$. Consequently, $(f \circ g)g' = f_{-1}g^{-1}g' + (F' \circ g)g' = f_{-1}g'/g + (F \circ g)'$ and (iv) follows from (i) and (iii). Property (v) is clear from the definition.

The Lagrange Inversion Formula

As mentioned above, any formal series $f \in K[[X]]$ with $f_0 = 0$ and $f_1 \neq 0$ has a composition inverse g in $K[[X]]$. The following relation between the coefficients of g^n and f^{-k} holds ("Lagrange inversion formula"):

$$k[X^k]g^n = n[X^{-n}]f^{-k}.$$

In particular, for $n = 1$ and all $k \geq 1$,

$$[X^k]g = \frac{1}{k}\mathrm{Res}\left(f^{-k}\right).$$

Since the proof of the Lagrange inversion formula is a very short computation, it is worth reporting it here. Since $\mathrm{ord}(f) = 1$, by the above rules of calculus,

$$\begin{aligned} k\left[X^k\right]g^n \quad &\aleph k\mathrm{Res}\left(g^n X^{-k-ü}\right) \quad k\mathrm{ord}(f)\mathrm{Res}\left(g^n X^{-k-}\right) \quad k\mathrm{Res}\left(((g^n X^{-k-}) \circ f) f'\right) \\ &= k\mathrm{Res}\left((g \circ f)^n f^{-k-1} f'\right) = k\mathrm{Res}\left(X^n f^{-k-1} f'\right) = -\mathrm{Res}\left(X^n (f^{-k})'\right) \\ &= \mathrm{Res}\left(\left(X^n\right)' f^{-k}\right) = n\mathrm{Res}\left(X^{n-1} f^{-k}\right) = n[X^{-n}]f^{-k}. \end{aligned}$$

Generalizations. One may observe that the above computation can be repeated plainly in more general settings than $K((X))$: a generalization of the Lagrange inversion formula is already available working in the $\mathrm{C}((X))$-modules $X^{\alpha}\mathrm{C}((X))$, where α is a complex exponent. As a consequence, if f and g are as above, with $f_1 = g_1 = 1$, we can relate the complex powers of f/X and g/X: precisely, if α and β are non-zero complex numbers with negative integer sum, $m=-\alpha-\beta \in \mathrm{N}$, then

$$\frac{1}{\alpha}[X^m]\left(\frac{f}{X}\right)^{\alpha} = -\frac{1}{\beta}[X^m]\left(\frac{g}{X}\right)^{\beta}.$$

For instance, this way one finds the power series for complex powers of the Lambert function.

Power Series in Several Variables

Formal power series in any number of indeterminates (even infinitely many) can be defined. If I is an index set and X_I is the set of indeterminates X_i for $i \in I$, then a monomial X^{α} is any finite product of elements of X_I (repetitions allowed); a formal power series in X_I with coefficients in a ring R is determined by any mapping from the set of monomials X^{α} to a corresponding coefficient c_{α}, and is denoted $\sum_{\alpha} c_{\alpha} X^{\alpha}$. The set of all such formal power series is denoted $R[[X_I]]$, and it is given a ring structure by defining,

$$\left(\sum_{\alpha} c_{\alpha} X^{\alpha}\right) + \left(\sum_{\alpha} d_{\alpha} X^{\alpha}\right) = \sum_{\alpha} (c_{\alpha} + d_{\alpha}) X^{\alpha}$$

and

$$\left(\sum_{\alpha} c_{\alpha} X^{\alpha}\right) \times \left(\sum_{\beta} d_{\beta} X^{\beta}\right) = \sum_{\alpha,\beta} c_{\alpha} d_{\beta} X^{\alpha+\beta}$$

Topology

The topology on $R[[X_I]]$ is such that a sequence of its elements converges only if for each monomial X^{α} the corresponding coefficient stabilizes. If I is finite, then this the J-adic topology, where J is the ideal of $R[[X_I]]$ generated by all the indeterminates in X_I. This does not hold if I is infinite. For example, if $I = \mathbb{N}$, then the sequence $(f_n)_{n\in\mathbb{N}}$ with $f_n = X_n + X_{n+1} + X_{n+2} + \ldots$ does not converge with respect to any J-adic topology on R, but clearly for each monomial the corresponding coefficient stabilizes.

As remarked above, the topology on a repeated formal power series ring like $R[[X]][[Y]]$ is usually chosen in such a way that it becomes isomorphic as a topological ring to $R[[X,Y]]$.

Operations

All of the operations defined for series in one variable may be extended to the several variables case.

- A series is invertible if and only if its constant term is invertible in R.
- The composition $f(g(X))$ of two series f and g is defined if f is a series in a single indeterminate, and the constant term of g is zero. For a series f in several indeterminates a form of "composition" can similarly be defined, with as many separate series in the place of g as there are indeterminates.

In the case of the formal derivative, there are now separate partial derivative operators, which differentiate with respect to each of the indeterminates. They all commute with each other.

Universal Property

In the several variables case, the universal property characterizing $R[[X_1, \ldots, X_r]]$ becomes the following. If S is a commutative associative algebra over R, if I is an ideal of S such that the I-adic topology on S is complete, and if $x_1, \ldots, x_r$ are elements of I, then there is a *unique* $\Phi : R[[X_1, \ldots, X_n]] \to S$ with the following properties:

- Φ is an R-algebra homomorphism
- Φ is continuous
- $\Phi(X_i) = x_i$ for $i = 1, \ldots, r$

Non-commuting Variables

The several variable case can be further generalised by taking *non-commuting variables* X_i for $i \in I$, where I is an index set and then a monomial X^α is any word in the X_I; a formal power series in X_I with coefficients in a ring R is determined by any mapping from the set of monomials X^α to a corresponding coefficient c_α, and is denoted $\sum_\alpha c_\alpha X^\alpha$. The set of all such formal power series is denoted $R«X_I»$, and it is given a ring structure by defining addition pointwise

$$\left(\sum_\alpha c_\alpha X^\alpha\right) + \left(\sum_\alpha d_\alpha X^\alpha\right) = \sum_\alpha (c_\alpha + d_\alpha) X^\alpha$$

and multiplication by,

$$\left(\sum_{\alpha} c_{\alpha} X^{\alpha}\right) \times \left(\sum_{\alpha} d_{\alpha} X^{\alpha}\right) = \sum_{\alpha,\beta} c_{\alpha} d_{\beta} X^{\alpha} \cdot X^{\beta}$$

where $\cdot$ denotes concatenation of words. These formal power series over R form the Magnus ring over R.

On a Semiring

In theoretical computer science, the following definition of a formal power series is given: let Σ be an alphabet (finite set) and S be a semiring. In this context, a formal power series is any mapping r from the set of strings generated by Σ (denoted as Σ^*) to the semiring S. The values of such a mapping r are (somewhat idiosyncratically) denoted as (r, w) were $w \in \Sigma^*$. Then the mapping r itself is conventionally written as $r = \sum_{w \in \Sigma^*} (r, w) w$. Given this notation, the values (r, w) are also called the coefficients of the series. Similarly to the non-commuting [ring] case discussed in the section above this, the notation for the collection of all power series given a fixed alphabet and semiring is $S \langle\langle \Sigma^* \rangle\rangle$.

Replacing the Index set by an Ordered Abelian Group

Suppose G is an ordered abelian group, meaning an abelian group with a total ordering respecting the group's addition, so that $a < b$ if and only if $a + c < b + c$ for all c. Let I be a well-ordered subset of G, meaning I contains no infinite descending chain. Consider the set consisting of,

$$\sum_{i \in I} a_i X^i$$

for all such I, with a_i in a commutative ring R, where we assume that for any index set, if all of the a_i are zero then the sum is zero. Then $R((G))$ is the ring of formal power series on G; because of the condition that the indexing set be well-ordered the product is well-defined, and we of course assume that two elements which differ by zero are the same. Sometimes the notation $[[R^G]]$ is used to denote $R((G))$.

Various properties of R transfer to $R((G))$. If R is a field, then so is $R((G))$. If R is an ordered field, we can order $R((G))$ by setting any element to have the same sign as its leading coefficient, defined as the least element of the index set I associated to a non-zero coefficient. Finally if G is a divisible group and R is a real closed field, then $R((G))$ is a real closed field, and if R is algebraically closed, then so is $R((G))$.

This theory is due to Hans Hahn, who also showed that one obtains subfields when the number of (non-zero) terms is bounded by some fixed infinite cardinality.

Examples and Related Topics

- Bell series are used to study the properties of multiplicative arithmetic functions.
- Formal groups are used to define an abstract group law using formal power series.

- Puiseux series are an extension of formal Laurent series, allowing fractional exponents.
- Rational series.

We will first try to develop the theory of generating functions by getting closed form expressions for some known recurrence relations. These ideas will be used later to get some binomial identities. $n \in \mathbb{Q}$ and $k \in \mathbb{Z}$, $k \geq 0$, the binomial coefficients, $\binom{n}{k}$, are well defined, using the idea that $\binom{n}{k} = \frac{n(n-1)(n-2)\cdots(n-k+1)}{k!}$. We now start with the definition of "formal power series" over $\mathbb{Q}$ and study its properties in some detail.

Definition: (Formal power series). An algebraic expression of the form $f(x) = \sum_{n\geq 0} a_n x^n$, where $a_n \in \mathbb{Q}$ for all n ≥ 0, is called a formal power series in the indeterminate x over $\mathbb{Q}$.

The set of all formal power series in the indeterminate x, with coefficients from $\mathbb{Q}$ will be denoted by $\mathcal{P}(x)$.

Remark: Given a sequence of numbers $\{a_n \in \mathbb{Q} : n = 0,1,2,\ldots\}$, one associates two formal power series, namely, $\sum_{n\geq 0} a_n x^n$ and $\sum_{n\geq 0} a_n \frac{x^n}{n!}$. The expression $\sum_{n\geq 0} a_n x^n$ is called the generating function and the expression $\sum_{n\geq 0} a_n \frac{x^n}{n!}$ is called the exponential generating function, for the numbers $\{a_n : n \geq 0\}$.

2. Let $f(x) = \sum_{n\geq 0} a_n x^n$ be a formal power series. Then the coefficient of x^n, for n ≥ 0, in $f(x)$ is denoted by $[x^n] f(x)$. That is, $a_0 = [x^0] f(x)$ and $a_n = [x^n] f(x)$, for n ≥ 1.

3. One thinks of $\sum_{n\geq 0} a_n x^n$ as an algebraic expression. In general, one is interested only in computing the coefficient of certain power of x and not in evaluating them for any value of x. But, if at all there is a need to evaluate it at a point, say x_0, then one needs to determine its "radius of convergence" and then evaluate it if x_0 lies within that radius.

We need the following definition to proceed further.

Definition: (Equality of two formal power series). Two elements $f(x) = \sum_{n\geq 0} a_n x^n$ and $g(x) = \sum_{n\geq 0} b_n x^n$ of $\mathcal{P}(x)$ are said to be equal if $a_n = b_n$, for all n ≥ 0.

We are now ready to define the algebraic rules:

Definition: Let $f(x) = \sum_{n\geq 0} a_n x^n, g(x) = \sum_{n\geq 0} b_n x^n \in \mathcal{P}(x)$. Then their

1. sum/addition is defined by

$$f(x) + g(x) = \sum_{n\geq 0} a_n x^n + \sum_{n\geq 0} b_n x^n = \sum_{n\geq 0} (a_n + b_n) x^n.$$

2. product is defined by

$$f(x)\cdot g(x)=\left(\sum_{n\geq 0}a_nx^n\right)\cdot\left(\sum_{n\geq 0}b_nx^n\right)=\sum_{n\geq 0}c_nx^n,\ where\ c_n=\sum_{k=0}^{n}a_kb_{n-k},\ for\ n\geq 0.$$

This product is also called the Cauchy product.

Remark: In case of exponential power series, the product of $f(x)=\sum_{n\geq 0}a_n\frac{x^n}{n!}$ and $g(x)=\sum_{n\geq 0}b_n\frac{x^n}{n!}$ equals $\sum_{n\geq 0}d_n\frac{x^n}{n!}$, where $d_n=\sum_{k=0}^{n}\binom{n}{k}a_kb_{n-k}\ for\ n\geq 0$

2. Note that the expression e^{e^x-1} is a well defined formal power series as the definition $e^y=\sum_{n\geq 0}\frac{y^n}{n!}$ implies that $e^{e^x-1}=\sum_{n\geq 0}\frac{(e^x-1)^n}{n!}$ and hence,

$$\left[x^m\right]e^{e^x-1}=\left[x^m\right]\sum_{n\geq 0}\frac{(e^x-1)^n}{n!}=\sum_{n=0}^{m}\left[x^m\right]\frac{(e^x-1)^n}{n!}.$$

That is, for each $m\geq 0,\left[x^m\right]e^{e^x-1}$ is a sum of a finite number of real numbers. Where as the expression e^{e^x} is not a formal power series as the computation of $\left[x^m\right]e^{e^x}$, for all m ≥ 0, will indeed require an infinite sum.

Thus, under the algebraic operations defined above, it can be checked that the set $\mathcal{P}(x)$ forms a Commutative Ring with identity, where the identity element is given by the formal power series $f(x)=1$. In this ring, the element $f(x)=\sum_{n\geq 0}a_nx^n$ is said to have a reciprocal if there exists an other element $g(x)=\sum_{n\geq 0}b_nx^n\in P(x)$ such that $f(x)\cdot g(x)=1$. So, the question arises, under what conditions on the coefficients of $f(x)$, can we find $g(x)\in\mathcal{P}(x)$ such that $f(x)g(x)=1$.

The answer to this question is given in the following proposition.

Proposition: Let $f(x)=\sum_{n\geq 0}a_nx^n\in\mathcal{P}(x)$. Then there exists $f(x)\in\mathcal{P}(x)$ satisfying $f(x)\cdot g(x)=1$ if and only if $a_0\neq 0$.

Proof. Let $g(x)=\sum_{n\geq 0}b_nx^n\in\mathcal{P}(x)$. Then, by the definition of Cauchy product, $f(x)g(x)=\sum_{n\geq 0}c_nx^n$ where $c_n=\sum_{n\geq 0}^{n}a_kb_{n-k}$, for all n ≥ 0. Therefore, using the definition of equality of two power series, we see that $f(x)g(x)=1$ if and only if $c_0 = 1$ and $c_n = 0$, for all n ≥ 1.

Therefore, if $a_0 = 0$ then $c_0 = 0$ and hence the Cauchy product $f(x)g(x)$ can never equal 1.

However, if $a_0\neq 0$, then the coefficients b_n's can be recursively obtained as follows:

$$b_0 = \frac{1}{a_0} \; as\, 1 = c_0 = a_0 b_0$$

$$b_1 = \frac{-1}{a_0} \cdot (a_1 b_0) \; as \; 0 = c_1 = a_0 b_1 + a_1 b_0.$$

$b_2 = \frac{-1}{a_0} \cdot (a_2 b_0 + a_1 b_1) \; as \; 0 = c_2 = a_0 b_2 + a_1 b_1 + a_2 b_0$. And in general, if we have already computed the values of b_k, for $k \le r$, then

$$b_{r+1} = \frac{-1}{a_0} \cdot (a_{r+1} b_0 + a_r b_1 + \cdots a_1 b_r) \; as \; 0 = c_{r+1} = a_{r+1} b_0 + a_r b_1 + \cdots + a_1 b_r + a_0 b_{r+1}.$$

Note that if the coefficients a_n's come from $\mathcal{R}$, a commutative ring with unity, then one needs bo to be an invertible element of $\mathcal{R}$ for Proposition 2.4.6 to hold true. Let us now look at the composition of two formal power series. Recall that, if $f(x) = \sum_{n \ge 0} a_n x^n, g(x) = \sum_{n \ge 0} b_n x^n \in \mathcal{P}(x)$ then the composition $((f \circ g)(x) = f(g(x)) = \sum_{n \ge 0} a_n (g(x))^n = \sum_{n \ge 0} a_n (\sum_{m \ge 0} b_m x^m)^n$ may not be defined (just to compute the constant term of the composition, one may have to look at an infinite sum). For example, let $f(x) = e^x$ and $g(x) = x + 1$. Note that $g(0) = 1 \ne 0$. Here, $(f \circ g)(x) = f(g(x)) = f(x+1) = e^{x+1}$. So, as function $f \circ g$ is well defined, but there is no formal procedure to write e^{x+1} as $\sum_{k \ge 0} a_k x^k \in \mathcal{P}(x)$ (i.e., with $a_k \in \mathbb{Q}$) and hence e^{x+1} is not a formal power series over $\mathbb{Q}$.

The next result gives the condition under which the composition $(f \circ g)(x)$ is well defined.

Proposition: Let $f(x) = \sum_{n \ge 0} a_n x^n$ and $g(x) = \sum_{n \ge 0} b_n x^n$ be two formal power series. Then the composition $(f \circ g)(x)$ is well defined if either f is a polynomial or $b_0 = 0$.

Moreover, suppose that $a_0 = 0$. Then, there exists $g(x) = \sum_{n \ge 0} b_n x^n$, with $b_0 = 0$, such that $(f \circ g)(x) = x$. Furthermore, $(g \circ f)(x)$ is well defined and $(g \circ f)(x) = x$.

Proof. Let $(f \circ g)(x) = f(g(x)) = \sum_{n \ge 0}^{n} c_n x^n$ and suppose that either f is a polynomial or $b_0 = 0$. Then to compute $c_k = [x^k](f \circ g)(x)$, for $k \ge 0$, one just needs to consider the terms $a_0 + a_1 g(x) + a_2 (g(x))^2 + \cdots + a_k (g(x))^k$. Hence, each c_k is a real number and $(f \circ g)(x)$ is well defined. This completes the proof of the first portion. The proof of the other part is left to the readers.

We now define the formal differentiation of elements of $\mathcal{P}(x)$ and state a result without proof.

Definition: (Differentiation). Let $f(x) = \sum_{n \ge 0} a_n x^n \in \mathcal{P}(x)$. Then the formal differentia-tion of $f(x)$, denoted $Df(x)$, is defined by,

$$Df(x) = a_1 + 2a_2 x + \cdots + n a_n x^{n-1} + \cdots = \sum_{n \ge 0} n a_n x^{n-1}.$$

Proposition: Let $f(x)=\sum_{n\geq 0} a_n x^n \in \mathcal{P}(x)$. Then $f(x)=a_0$, a constant, whenever $Df(x)=0$. Also, $f(x)=a_0 e^x$ whenever $Df(x)=f(x)$.

Before proceeding further, let us look at some important examples, where the above results can be used.

Example: Using the generalized binomial theorem, recall that, for a positive integer r, the formal power series,

$$\sum_{n\geq 0}\binom{n+r-1}{n}x^n \textit{ equals } \frac{1}{(1-x)^r}.$$

In particular, for r = 1, the formal power series $\sum_{n\geq 0} nx^n$ equals $\frac{1}{(1-x)}$.

1. Determine a closed form expression for $\sum_{n\geq 0} nx^n \in \mathcal{P}(x)$.

Solution: As $\frac{1}{1-x}=\sum_{n\geq 0} x^n$, *one has* $\frac{1}{(1-x)^2}=D\left(\frac{1}{1-x}\right)=D\left(\sum_{n\geq 0} x^n\right)=\sum_{n\geq 0} nx^{n-1}$.

Thus, the closed form expression is $\frac{1}{(1-x)^2}$.

This can also be computed as follows:

Let $S=\sum_{n\geq 0} nx^n = x+2x^2+3x^3+\cdots$. Then $xS=x^2+2x^3+3x^4+\cdots$. Hence,

$(1-x)S=x+x^2+x^3+\cdots = x(1+x+x^2+\cdots)=\frac{x}{1-x}$. Thus, $S=\frac{1}{(1-x)^2}$.

2. Let $f(x)=\sum_{n\geq 0} a_n x^n \in P(x)$. Determine $\sum_{k\geq 0}^{n} a_k$.

Solution: Recall that the Cauchy product of $f(x)=\sum_{n\geq 0} a_n x^n$ and $g(x)=\sum_{n\geq 0} b_n x^n$ equals $\sum_{n\geq 0} c_n x^n$ where $c_n=\sum_{k=0}^{n} a_k b_{n-k}$, for n ≥ 0. Therefore, to get $c_n=\sum_{k=0}^{n} a_k$, one needs $b_k = 1$, for all k ≥ 0. That is, $\sum_{k=0}^{n} a_k = c_n\left[x^n\right]\left(f(x).\frac{1}{1-x}\right)$.

Hence, the Cauchy product helps us in computing the sum of the first N coefficients of a formal power series, for any N ≥ 1.

3. Determine the sum of the squares of the first N positive integers.

Solution: Using Equation (1), observe that $k=\left[x^{k-1}\right]\left(\frac{1}{(1-x)^2}\right)$. Therefore, using example, one has

$$\sum_{k=1}^{N} k = \left[x^{N-1}\right]\left(\frac{1}{(1-x)^2}\cdot\frac{1}{1-x}\right) = \left[x^{N-1}\right]\frac{1}{(1-x)^3} = \binom{N-1+3-1}{N-1} = \frac{N(N+1)}{2}.$$

4. Determine a closed form expression for $\sum_{k=1}^{N} k^2$.

Solution: Using example, observe that $\sum_{k\geq 0} kx^k = \frac{1}{(1-x)^2}$. Therefore, using the differentiation operator, one obtains:

$$\sum_{k\geq 0} k^3x^k = x\left(\sum_{k\geq 0} k^2x^{k-1}\right) = xD\left(\frac{x}{(1-x)^2}\right) = \frac{x(1+x)}{(1-x)^3}.$$

Thus, by Example:

$$\begin{aligned}\sum_{k=1}^{N} k^2 &= \left[x^N\right]\left(\frac{x(1+x)}{(1-x)^3}\cdot\frac{1}{1-x}\right) = \left[x^{N-1}\right]\left(\frac{1}{(1-x)^4}\right) + \left[x^{N-2}\right]\left(\frac{1}{(1-x)^4}\right)\\ &= \binom{N-1+4-1}{N-1} + \binom{N-2+4-1}{N-2}\\ &= \frac{N(N+1)(2N+1)}{6}.\end{aligned}$$

5. Determine a closed form expression for $\sum_{k=1}^{N} k^3$.

Solution: Using Equation (2), observe that $\sum_{k\geq 0} k^2x^k = \frac{x(1+x)}{(1-x)^3}$. So,

$$\sum_{k\geq 0} k^2x^k = x\left(\sum_{k\geq 0} k^2x^{k-1}\right) = xD\left(\frac{x(1+x)}{(1-x)^3}\right) = \frac{x(1+4x+x^2)}{(1-x)^4}.$$

Thus, by Example:

$$\begin{aligned}\sum_{k=1}^{N} k^3 &= \left[x^N\right]\left(\frac{x(1+4x+x^2)}{(1-x)^4)}\cdot\frac{1}{1-x}\right)\\ &= \left[x^{N-1}\right]\left(\frac{1}{(1-x)^5}\right) + \left[x^{N-2}\right]\left(\frac{4}{(1-x)^5}\right) + \left[x^{N-3}\right]\left(\frac{1}{(1-x)^5}\right)\\ &= \binom{N-1+5-1}{N-1} + 4\binom{N-2+5-1}{N-2} + \binom{N-3+5-1}{N-3}\\ &= \left(\frac{N(N+1)}{2}\right)^2.\end{aligned}$$

Hence, we observe that we can inductively use this technique to get a closed form expression for $\sum_{k=1}^{N} k^r$, for any positive integer r.

6. Determine a closed form expression for $\sum_{n\geq 0}\frac{n^2+n+6}{n!}$.

Solution: As we need to compute the infinite sum, Cauchy product cannot be used. Also, one needs to find a convergent series, which when evaluated at some x_0, gives the required expression. Therefore, recall that the series $e^x=\sum_{n\geq 0}\frac{x^n}{n!}$, converges for all $x\in\mathbb{R}$ and evaluating e^x at $x=1$, gives $\sum_{n\geq 0}\frac{1}{n!}=e.$ Similarly, $\frac{n^2}{n!}=[x^n](xD(xDe^x))=[x^n]((x+x^2)e^x)$ and $\frac{n}{n!}=[x^n](xD(xDe^x))=[x^n]((x+x^2)e^x)$. Thus,

$$\sum_{n\geq 0}=\frac{n^2+n+6}{n!}=(x+x^2)e^x+xe^x+6e^x\ \Big|_{x=1}=9e.$$

7. Let n and r be two fixed positive integers. Then determine the number of non-negative integer solutions to the system $x_1+x_2+\cdots+x_n=r$.

Solution: Recall that this number equals $\binom{r+n-1}{r}$

We can think of the problem as follows: the above system can be interpreted as coming from the monomial x^r, where $r=x_1+x_2+\cdots+x_n$. That is, the problem reduces to finding the coefficients of y^{x_k} of a formal power series, for non-negative integers x_k's.

Now, recall that the terms y^{x_k} appear with a coefficient 1 in the expression $\frac{1}{1-y}=\sum_{i\geq 0}y^i$.

Hence, the question reduces to computing:

$$[y^r]=\left(\frac{1}{(1-y)(1-y)\cdots(1-y)}\right)=[y^r]\frac{1}{(1-y)^n}=\binom{r+n-1}{r}.$$

We now look at some examples that may require the use of the package "MATHEMATICA" or "MAPLE" or ... to obtain the exact answer. So, in the examples given below, one is interested in getting only a formal power series whose coefficient gives the required result even though it may be difficult to compute the coefficient.

Example: For fixed positive integers n and r, determine the number of non-negative integer solutions to the system $x_1+2x_2+\cdots+nx_n=r$.

Solution: The ideas from example imply that one needs to consider the formal power series $\sum_{i\geq 0}x^{ki}=\frac{1}{1-x^k}$. Hence, need to compute:

$$[x^r]\frac{1}{(1-x)(1-x^2)\cdots(1-x^n)}.$$

2. Determine the number of solutions to the system $x_1+x_2+\cdots+x_5=n$, where x_i's are non-negative integer, $x_1\geq 4$, $x_4\leq 10$ and x_r is a multiple of r, whenever $r\neq 1,4$.

Solution: The condition $x_1 \geq 4$ corresponds to looking at x^k, for k ≥ 4, which corresponds to the formal power series $\sum_{k\geq 4} x^k$. Similarly, $x_4 \leq 10$ gives the formal power series $\sum_{k=0}^{10} x^k$ and the condition x_r is a multiple of r, for $r \neq 1,4$, corresponds to $\sum_{k\geq 0} x^{rk}$. So, we need to compute the coefficient of x^n in the product

$$\left(\sum_{k\geq 4} x^k\right)\cdot\left(\sum_{k=0}^{10} x^k\right)\cdot\left(\sum_{k\geq 0} x^{2k}\right)\cdot\left(\sum_{k\geq 0} x^{3k}\right)\cdot\left(\sum_{k\geq 0} x^{5k}\right) = \frac{x^4(1-x^{11})}{(1-x)^2(1-x^2)(1-x^3)(1-x^5)}.$$

3. Determine the number of ways in which 100 voters can cast their 100 votes for 10 candidates such that no candidate gets more than 20 votes.

Solution: Note that we are assuming that the voters are identical. So, we need to solve the system in non-negative integers to the system $x_1 + x_2 + \cdots + x_{10} = 100$, with $0 \leq x_i \leq 20$, for $1 \leq i \leq 10$. So, we need to find the coefficient of x^{100} in,

$$\left(\sum_{k=1}^{20} x^k\right)^{10} = \frac{(1-x^{21})^{10}}{(1-x)^{10}} = \left(\sum_{i=0}^{10} (-1)^i \binom{10}{i} x^{21i}\right)\cdot\left(\sum_{j\geq 0} \binom{10+j-1}{j} x^j\right).$$

Thus, the required answer equals $[x^{100}]\left(\sum_{k=1}^{20} x^k\right)^{10} = \sum_{i=0}^{4} (-1)^i \binom{10}{i}\cdot\binom{109-21i}{9}$.

Before moving to the applications of formal power series to the solution of recurrence rela- tions, let us list a few well known power series. The readers are requested to get a proof for their satisfaction.

Table of Formal Power Series,

e^x	$= \sum_{k\geq 0} \frac{x^k}{k!}$	$\log(1\text{-}x)$	$= -\sum_{k\geq 1} \frac{x^k}{k}, \|x\|<1$
$(1+x)^a$	$= \sum_{r\geq 0} \binom{a}{k} x^k, \|x\|<1$	$\frac{1}{1-x}$	$= \sum_{k\geq 0} x^k, \|x\|<1$
$\frac{1}{(1-x)^a}$	$= \sum_{k\geq 0} \binom{a+k-1}{k} x^k, \|x\|<1$	$\frac{1}{\sqrt{1-4x}}$	$= \sum_{k\geq 0} \binom{2k}{k} x^k, \|x\|<\frac{1}{4}$
$x^{-r}(1+x)^n$	$= \sum_{k\geq -r} \binom{n}{r+k} x^k, \|x\|<1$	$\frac{x^n}{(1-x)^{n+1}}$	$= \sum_{k\geq 0} \binom{k}{n} x^k, n\geq 0, \|x\|<1$
$\sin(x)$	$= \sum_{r\geq 0} \frac{(-1)^r x^{2r+1}}{(2r+1)!}$	$\cos(x)$	$= \sum_{r\geq 0} \frac{(-1)^r x^{2r}}{(2r)!}$
$\sinh(x)$	$= \sum_{r\geq 0} \frac{x^{2r+1}}{(2r+1)!}$	$\cosh(x)$	$= \sum_{r\geq 0} \frac{x^{2r}}{(2r)!}$

$$\frac{1-\sqrt{1-4x}}{2x} = \sum_{k\geq 0} \frac{1}{k+1}\binom{2k}{k} x^k, |x|<\frac{1}{4}$$

Recurrence Relation

In mathematics, a recurrence relation is an equation that recursively defines a sequence or multidimensional array of values, once one or more initial terms are given: each further term of the sequence or array is defined as a function of the preceding terms.

The term difference equation sometimes refers to a specific type of recurrence relation. However, "difference equation" is frequently used to refer to *any* recurrence relation.

Examples

Logistic Map

An example of a recurrence relation is the logistic map:

$$x_{n+1} = rx_n(1-x_n),$$

with a given constant r; given the initial term x_0 each subsequent term is determined by this relation.

Some simply defined recurrence relations can have very complex (chaotic) behaviours, and they are a part of the field of mathematics known as nonlinear analysis.

Solving a recurrence relation means obtaining a closed-form solution: a non-recursive function of n.

Fibonacci Numbers

The recurrence satisfied by the Fibonacci numbers is the archetype of a homogeneous linear recurrence relation with constant coefficients. The Fibonacci sequence is defined using the recurrence

$$F_n = F_{n-1} + F_{n-2}$$

with seed values,

$$F_0 = 0$$
$$F_1 = 1.$$

Explicitly, the recurrence yields the equations,

$$F_2 = F_1 + F_0$$
$$F_3 = F_2 + F_1$$
$$F_4 = F_3 + F_2 \text{ etc.}$$

We obtain the sequence of Fibonacci numbers, which begins:

0, 1, 1, 2, 3, 5, 8, 13, 21, 34, 55, 89, ...

The recurrence can be solved by methods described below yielding Binet's formula, which involves powers of the two roots of the characteristic polynomial $t^2 = t + 1$; the generating function of the sequence is the rational function,

$$\frac{t}{1-t-t^2}.$$

Binomial Coefficients

A simple example of a multidimensional recurrence relation is given by the binomial coefficients $\binom{n}{k}$, which count the number of ways of selecting k elements out of a set of n elements. They can be computed by the recurrence relation,

$$\binom{n}{k} = \binom{n-1}{k-1} + \binom{n-1}{k},$$

with the base cases $\binom{n}{0} = \binom{n}{n} = 1$. Using this formula to compute the values of all binomial coefficients generates an infinite array called Pascal's triangle. The same values can also be computed directly by a different formula that is not a recurrence, but that requires multiplication and not just addition to compute:

$$\binom{n}{k} = \frac{n!}{k!(n-k)!}.$$

Relationship to Difference Equations Narrowly Defined

Given an ordered sequence $\{a_n\}_{n=1}^{\infty}$ of real numbers: the first difference $\Delta(a_n)$ is defined as

$$\Delta(a_n) = a_{n+1} - a_n$$

The second difference $\Delta^2(a_n)$ is defined as,

$$\Delta^2(a_n) = \Delta(a_{n+1}) - \Delta(a_n),$$

which can be simplified to,

$$\Delta^2(a_n) = a_{n+2} - 2a_{n+1} + a_n$$

More generally: the kth difference of the sequence a_n written as $\Delta^k(a_n)$ is defined recursively as,

$$\Delta^k(a_n) = \Delta^{k-1}(a_{n+1}) - \Delta^{k-1}(a_n) = \sum_{t=0}^{k} \binom{k}{t} (-1)^t a_{n+k-t}$$

(The sequence and its differences are related by a binomial transform.) The more restrictive definition of difference equation is an equation composed of a_n and its kth differences. (A widely used broader definition treats "difference equation" as synonymous with "recurrence relation".)

Actually, it is easily seen that,

$$a_{n+k} = \binom{k}{0} a_n + \binom{k}{1} \Delta(a_n) + \cdots + \binom{k}{k} \Delta^k(a_n)$$

Thus, a difference equation can be defined as an equation that involves a_n, a_{n-1}, a_{n-2} etc. (or equivalenty a_n, a_{n+1}, a_{n+2} etc.)

Since difference equations are a very common form of recurrence, some authors use the two terms interchangeably. For example, the difference equation:

$$3\Delta^2(a_n) + 2\Delta(a_n) + 7a_n = 0$$

is equivalent to the recurrence relation,

$$3a_{n+2} = 4a_{n+1} - 8a_n$$

Thus one can solve many recurrence relations by rephrasing them as difference equations, and then solving the difference equation, analogously to how one solves ordinary differential equations. However, the Ackermann numbers are an example of a recurrence relation that do not map to a difference equation, much less points on the solution to a differential equation.

Summation equations relate to difference equations as integral equations relate to differential equations.

From Sequences to Grids

Single-variable or one-dimensional recurrence relations are about sequences (i.e. functions defined on one-dimensional grids). Multi-variable or n-dimensional recurrence relations are about n-dimensional grids. Functions defined on n-grids can also be studied with partial difference equations.

Solving

Roots of the Characteristic Polynomial

An order-d homogeneous linear recurrence with constant coefficients is an equation of the form

$$a_n = c_1 a_{n-1} + c_2 a_{n-2} + \cdots + c_d a_{n-d},$$

where the d coefficients c_i (for all i) are constants.

A constant-recursive sequence is a sequence satisfying a recurrence of this form. There are d degrees of freedom for solutions to this recurrence, i.e., the initial values $a_0, \ldots, a_{d-1}$ can be taken to be any values but then the recurrence determines the sequence uniquely.

The same coefficients yield the characteristic polynomial (also "auxiliary polynomial"):

$$p(t) = t^d - c_1 t^{d-1} - c_2 t^{d-2} - \cdots - c_d$$

whose d roots play a crucial role in finding and understanding the sequences satisfying the recurrence. If the roots $r_1, r_2, \ldots$ are all distinct, then each solution to the recurrence takes the form:

$$a_n = k_1 r_1^n + k_2 r_2^n + \cdots + k_d r_d^n,$$

where the coefficients k_i are determined in order to fit the initial conditions of the recurrence. When the same roots occur multiple times, the terms in this formula corresponding to the second and later occurrences of the same root are multiplied by increasing powers of n. For instance, if the characteristic polynomial can be factored as $(x-r)^3$, with the same root r occurring three times, then the solution would take the form:

$$a_n = k_1 r^n + k_2 n r^n + k_3 n^2 r^n.$$

As well as the Fibonacci numbers, other constant-recursive sequences include the Lucas numbers and Lucas sequences, the Jacobsthal numbers, the Pell numbers and more generally the solutions to Pell's equation.

For order 1, the recurrence:

$$a_n = r a_{n-1}$$

has the solution $a_n = r^n$ with $a_0 = 1$ and the most general solution is $a_n = kr^n$ with $a_0 = k$. The characteristic polynomial equated to zero (the characteristic equation) is simply $t - r = 0$.

Solutions to such recurrence relations of higher order are found by systematic means, often using the fact that $a_n = r^n$ is a solution for the recurrence exactly when $t = r$ is a root of the characteristic polynomial. This can be approached directly or using generating functions (formal power series) or matrices.

Consider, for example, a recurrence relation of the form:

$$a_n = A a_{n-1} + B a_{n-2}.$$

When does it have a solution of the same general form as $a_n = r^n$? Substituting this guess (ansatz) in the recurrence relation, we find that:

$$r^n = A r^{n-1} + B r^{n-2}$$

must be true for all $n > 1$.

Dividing through by r^{n-2}, we get that all these equations reduce to the same thing;

$$r^2 = Ar + B,$$
$$r^2 - Ar - B = 0,$$

which is the characteristic equation of the recurrence relation. Solve for r to obtain the two roots λ_1, λ_2: these roots are known as the characteristic roots or eigenvalues of the characteristic equation. Different solutions are obtained depending on the nature of the roots: If these roots are distinct, we have the general solution,

$$a_n = C\lambda_1^n + D\lambda_2^n$$

while if they are identical (when $A^2 + 4B = 0$), we have:

$$a_n = C\lambda^n + Dn\lambda^n$$

This is the most general solution; the two constants C and D can be chosen based on two given initial conditions a_0 and a_1 to produce a specific solution.

In the case of complex eigenvalues (which also gives rise to complex values for the solution parameters C and D), the use of complex numbers can be eliminated by rewriting the solution in trigonometric form. In this case we can write the eigenvalues as $\lambda_1, \lambda_2 = \alpha \pm \beta i$. Then it can be shown that

$$a_n = C\lambda_1^n + D\lambda_2^n.$$

can be rewritten as,

$$a_n = 2M^n \left(E\cos(\theta n) + F\sin(\theta n)\right) = 2GM^n \cos(\theta n - \delta),$$

where

$$M = \sqrt{\alpha^2 + \beta^2} \quad \cos(\theta) = \tfrac{\alpha}{M} \quad \sin(\theta) = \tfrac{\beta}{M}$$
$$C, D = E \mp Fi$$
$$G = \sqrt{E^2 + F^2} \quad \cos(\delta) = \tfrac{E}{G} \quad \sin(\delta) = \tfrac{F}{G}$$

Here E and F (or equivalently, G and δ) are real constants which depend on the initial conditions. Using

$$\lambda_1 + \lambda_2 = 2\alpha = A,$$
$$\lambda_1 \cdot \lambda_2 = \alpha^2 + \beta^2 = -B,$$

one may simplify the solution given above as,

$$a_n = (-B)^{\frac{n}{2}} \left(E\cos(\theta n) + F\sin(\theta n)\right),$$

where a_1 and a_2 are the initial conditions and

$$E = \frac{-Aa_1 + a_2}{B}$$

$$F = -i\frac{A^2 a_1 - Aa_2 + 2a_1 B}{B\sqrt{A^2 + 4B}}$$

$$\theta = \arccos\left(\frac{A}{2\sqrt{-B}}\right)$$

In this way there is no need to solve for λ_1 and λ_2.

In all cases—real distinct eigenvalues, real duplicated eigenvalues, and complex conjugate eigenvalues—the equation is stable (that is, the variable a converges to a fixed value [specifically, zero]) if and only if *both* eigenvalues are smaller than one in absolute value. In this second-order case, this condition on the eigenvalues can be shown to be equivalent to $|A| < 1 - B < 2$, which is equivalent to $|B| < 1$ and $|A| < 1 - B$.

The equation in the above example was homogeneous, in that there was no constant term. If one starts with the non-homogeneous recurrence

$$b_n = Ab_{n-1} + Bb_{n-2} + K$$

with constant term K, this can be converted into homogeneous form as follows: The steady state is found by setting $b_n = b_{n-1} = b_{n-2} = b^*$ to obtain,

$$b^* = \frac{K}{1 - A - B}.$$

Then the non-homogeneous recurrence can be rewritten in homogeneous form as,

$$[b_n - b^*] = A[b_{n-1} - b^*] + B[b_{n-2} - b^*],$$

which can be solved as above.

The stability condition stated above in terms of eigenvalues for the second-order case remains valid for the general n^{th}-order case: the equation is stable if and only if all eigenvalues of the characteristic equation are less than one in absolute value.

Given a homogeneous linear recurrence relation with constant coefficients of order d, let $p(t)$ be the characteristic polynomial (also "auxiliary polynomial"),

$$t^d - c_1 t^{d-1} - c_2 t^{d-2} - \cdots - c_d = 0$$

such that each c_i corresponds to each c_i in the original recurrence relation. Suppose λ is a root of $p(t)$ having multiplicity r. This is to say that $(t-\lambda)^r$ divides $p(t)$. The following two properties hold:

1. Each of the r sequences $\lambda^n, n\lambda^n, n^2\lambda^n, \ldots, n^{r-1}\lambda^n$ satisfies the recurrence relation.

2. Any sequence satisfying the recurrence relation can be written uniquely as a linear combination of solutions constructed in part 1 as λ varies over all distinct roots of $p(t)$.

As a result of this theorem a homogeneous linear recurrence relation with constant coefficients can be solved in the following manner:

1. Find the characteristic polynomial $p(t)$.
2. Find the roots of $p(t)$ counting multiplicity.
3. Write a_n as a linear combination of all the roots (counting multiplicity as shown in the theorem above) with unknown coefficients b_i.

$$a_n = \left(b_1\lambda_1^n + b_2 n\lambda_1^n + b_3 n^2\lambda_1^n + \cdots + b_r n^{r-1}\lambda_1^n\right) + \cdots + \left(b_{d-q+1}\lambda_*^n + \cdots + b_d n^{q-1}\lambda_*^n\right)$$

This is the general solution to the original recurrence relation. (q is the multiplicity of λ_*)

4. Equate each $a_0, a_1, \ldots, a_d$ from part 3 (plugging in n = 0, ..., d into the general solution of the recurrence relation) with the known values $a_0, a_1, \ldots, a_d$ from the original recurrence relation. However, the values a_n from the original recurrence relation used do not usually have to be contiguous: excluding exceptional cases, just d of them are needed (i.e., for an original homogeneous linear recurrence relation of order 3 one could use the values a_0, a_1, a_4). This process will produce a linear system of d equations with d unknowns. Solving these equations for the unknown coefficients $b_1, b_2, \ldots, b_d$ of the general solution and plugging these values back into the general solution will produce the particular solution to the original recurrence relation that fits the original recurrence relation's initial conditions (as well as all subsequent values $a_0, a_1, a_2, \ldots$ of the original recurrence relation).

The method for solving linear differential equations is similar to the method above—the "intelligent guess" (ansatz) for linear differential equations with constant coefficients is $e^{\lambda x}$ where λ is a complex number that is determined by substituting the guess into the differential equation.

This is not a coincidence. Considering the Taylor series of the solution to a linear differential equation:

$$\sum_{n=0}^{\infty} \frac{f^{(n)}(a)}{n!}(x-a)^n$$

it can be seen that the coefficients of the series are given by the nth derivative of $f(x)$ evaluated at the point a. The differential equation provides a linear difference equation relating these coefficients.

This equivalence can be used to quickly solve for the recurrence relationship for the coefficients in the power series solution of a linear differential equation.

The rule of thumb (for equations in which the polynomial multiplying the first term is non-zero at zero) is that:

$$y^{[k]} \rightarrow f[n+k]$$

and more generally

$$x^m * y^{[k]} \to n(n-1)...(n-m+1)f[n+k-m]$$

Example: The recurrence relationship for the Taylor series coefficients of the equation:

$$(x^2+3x-4)y^{[3]}-(3x+1)y^{[2]}+2y=0$$

is given by

$$n(n-1)f[n+1]+3nf[n+2]-4f[n+3]-3nf[n+1]-f[n+2]+2f[n]=0$$

or

$$-4f[n+3]+2nf[n+2]+n(n-4)f[n+1]+2f[n]=0.$$

This example shows how problems generally solved using the power series solution method taught in normal differential equation classes can be solved in a much easier way.

Example: The differential equation,

$$ay''+by'+cy=0$$

has solution,

$$y=e^{ax}.$$

The conversion of the differential equation to a difference equation of the Taylor coefficients is,

$$af[n+2]+bf[n+1]+cf[n]=0.$$

It is easy to see that the nth derivative of e^{ax} evaluated at 0 is a^n

Solving via Linear Algebra

A linearly recursive sequence y of order n,

$$y_{n+k}-c_{n-1}y_{n-1+k}-c_{n-2}y_{n-2+k}+\cdots-c_0y_k=0$$

is identical to,

$$y_n=c_{n-1}y_{n-1}+c_{n-2}y_{n-2}+\cdots+c_0y_0.$$

Expanded with n-1 identities of kind $y_{n-k}=y_{n-k}$, this n-th order equation is translated into a matrix difference equation system of n first-order linear equations,

$$\vec{y}_n = \begin{bmatrix} y_n \\ y_{n-1} \\ \vdots \\ \vdots \\ y_1 \end{bmatrix} = \begin{bmatrix} c_{n-1} & c_{n-2} & \cdots & \cdots & c_0 \\ 1 & 0 & \cdots & \cdots & 0 \\ 0 & \ddots & \ddots & & \vdots \\ \vdots & \ddots & \ddots & \ddots & \vdots \\ 0 & \cdots & 0 & 1 & 0 \end{bmatrix} \begin{bmatrix} y_{n-1} \\ y_{n-2} \\ \vdots \\ \vdots \\ y_0 \end{bmatrix} = C\,\vec{y}_{n-1} = C^n \vec{y}_0.$$

Observe that the vector $\vec{y}_n$ can be computed by n applications of the companion matrix, C, to the initial state vector, y_0. Thereby, n-th entry of the sought sequence y, is the top component of $\vec{y}_n, y_n = \vec{y}_n[n]$.

Eigendecomposition, $\vec{y}_n = C^n \vec{y}_0 = c_1 \lambda_1^n \vec{e}_1 + c_2 \lambda_2^n \vec{e}_2 + \cdots + c_n \lambda_n^n \vec{e}_n$ into eigenvalues, $\lambda_1, \lambda_2, \ldots, \lambda_n$, and eigenvectors, $\vec{e}_1, \vec{e}_2, \ldots, \vec{e}_n$, is used to compute $\vec{y}_n$.Thanks to the crucial fact that system C time-shifts every eigenvector, e, by simply scaling its components λ times,

$$C\vec{e}_i = \lambda_i \vec{e}_i = C \begin{bmatrix} e_{i,n} \\ e_{i,n-1} \\ \vdots \\ e_{i,1} \end{bmatrix} = \begin{bmatrix} \lambda_i e_{i,n} \\ \lambda_i e_{i,n-1} \\ \vdots \\ \lambda_i e_{i,1} \end{bmatrix}$$

that is, time-shifted version of eigenvector,e, has components λ times larger, the eigenvector components are powers of λ, $\vec{e}_i = \begin{bmatrix} \lambda_i^{n-1} & \cdots & \lambda_i^2 & \lambda_i & 1 \end{bmatrix}^T$, and, thus, recurrent homogeneous linear equation solution is a combination of exponential functions, $\vec{y}_n = \sum_1^n c_i \lambda_i^n \vec{e}_i$. The components c_i can be determined out of initial conditions:

$$\vec{y}_0 = \begin{bmatrix} y_0 \\ y_{-1} \\ \vdots \\ y_{-n+1} \end{bmatrix} = \sum_{i=1}^{n} c_i \lambda_i^0 \vec{e}_i = [\vec{e}_1 \quad \vec{e}_2 \quad \cdots \quad \vec{e}_n] \begin{bmatrix} c_1 \\ c_2 \\ \cdots \\ c_n \end{bmatrix} = E \begin{bmatrix} c_1 \\ c_2 \\ \cdots \\ c_n \end{bmatrix}$$

Solving for coefficients,

$$\begin{bmatrix} c_1 \\ c_2 \\ \cdots \\ c_n \end{bmatrix} = E^{-1} \vec{y}_0 = \begin{bmatrix} \lambda_1^{n-1} & \lambda_2^{n-1} & \cdots & \lambda_n^{n-1} \\ \vdots & \vdots & \ddots & \vdots \\ \lambda_1 & \lambda_2 & \cdots & \lambda_n \\ 1 & 1 & \cdots & 1 \end{bmatrix}^{-1} \begin{bmatrix} y_0 \\ y_{-1} \\ \vdots \\ y_{-n+1} \end{bmatrix}.$$

This also works with arbitrary boundary conditions $\underbrace{y_a, y_b, \ldots}_{n}$, not necessary the initial ones,

$$\begin{bmatrix} y_a \\ y_b \\ \vdots \end{bmatrix} = \begin{bmatrix} \vec{y}_a[n] \\ \vec{y}_b[n] \\ \vdots \end{bmatrix} = \begin{bmatrix} \sum_{i=1}^{n} c_i \lambda_i^a \vec{e}_i[n] \\ \sum_{i=1}^{n} c_i \lambda_i^b \vec{e}_i[n] \\ \vdots \end{bmatrix} = \begin{bmatrix} \sum_{i=1}^{n} c_i \lambda_i^a \lambda_i^{n-1} \\ \sum_{i=1}^{n} c_i \lambda_i^b \lambda_i^{n-1} \\ \vdots \end{bmatrix} =$$

$$= \begin{bmatrix} \sum c_i \lambda_i^{a+n-1} \\ \sum c_i \lambda_i^{b+n-1} \\ \vdots \end{bmatrix} = \begin{bmatrix} \lambda_1^{a+n-1} & \lambda_2^{a+n-1} & \cdots & \lambda_n^{a+n-1} \\ \lambda_1^{b+n-1} & \lambda_2^{b+n-1} & \cdots & \lambda_n^{b+n-1} \\ \vdots & \vdots & \ddots & \vdots \end{bmatrix} \begin{bmatrix} c_1 \\ c_2 \\ \vdots \\ c_n \end{bmatrix}.$$

This description is really no different from general method above, however it is more succinct. It also works nicely for situations like,

$$\begin{cases} a_n = a_{n-1} - b_{n-1} \\ b_n = 2a_{n-1} + b_{n-1}. \end{cases}$$

where there are several linked recurrences.

Solving with z-transforms

Certain difference equations - in particular, linear constant coefficient difference equations - can be solved using z-transforms. The z-transforms are a class of integral transforms that lead to more convenient algebraic manipulations and more straightforward solutions. There are cases in which obtaining a direct solution would be all but impossible, yet solving the problem via a thoughtfully chosen integral transform is straightforward.

Solving Non-homogeneous Linear Recurrence Relations with Constant Coefficients

If the recurrence is non-homogeneous, a particular solution can be found by the method of undetermined coefficients and the solution is the sum of the solution of the homogeneous and the particular solutions. Another method to solve an non-homogeneous recurrence is the method of *symbolic differentiation*. For example, consider the following recurrence:

$$a_{n+1} = a_n + 1$$

This is an non-homogeneous recurrence. If we substitute $n \mapsto n+1$, we obtain the recurrence:

$$a_{n+2} = a_{n+1} + 1$$

Subtracting the original recurrence from this equation yields,

$$a_{n+2} - a_{n+1} = a_{n+1} - a_n$$

or equivalently,

$$a_{n+2} = 2a_{n+1} - a_n$$

This is a homogeneous recurrence, which can be solved by the methods explained above. In general, if a linear recurrence has the form,

$$a_{n+k} = \lambda_{k-1}a_{n+k-1} + \lambda_{k-2}a_{n+k-2} + \cdots + \lambda_1 a_{n+1} + \lambda_0 a_n + p(n)$$

where $\lambda_0, \lambda_1, \ldots, \lambda_{k-1}$ are constant coefficients and $p(n)$ is the inhomogeneity, then if $p(n)$ is a polynomial with degree r, then this non-homogeneous recurrence can be reduced to a homogeneous recurrence by applying the method of symbolic differencing r times.

If

$$P(x) = \sum_{n=0}^{\infty} p_n x^n$$

is the generating function of the inhomogeneity, the generating function:

$$A(x) = \sum_{n=0}^{\infty} a(n) x^n$$

of the non-homogeneous recurrence,

$$a_n = \sum_{i=1}^{s} c_i a_{n-i} + p_n, \quad n \geq n_r,$$

with constant coefficients c_i is derived from,

$$\left(1 - \sum_{i=1}^{s} c_i x^i\right) A(x) = P(x) + \sum_{n=0}^{n_r-1} [a_n - p_n] x^n - \sum_{i=1}^{s} c_i x^i \sum_{n=0}^{n_r-i-1} a_n x^n.$$

If $P(x)$ is a rational generating function, $A(x)$ is also one. The case discussed above, where $p_n = K$ is a constant, emerges as one example of this formula, with $P(x) = K/(1-x)$. Another example, the recurrence $a_n = 10a_{n-1} + n$ with linear inhomogeneity, arises in the definition of the schizophrenic numbers. The solution of homogeneous recurrences is incorporated as $p = P = 0$.

Solving First-order Non-homogeneous Recurrence Relations with Variable Coefficients

Moreover, for the general first-order non-homogeneous linear recurrence relation with variable coefficients;

$$a_{n+1} = f_n a_n + g_n, \qquad f_n \neq 0,$$

there is also a nice method to solve it:

$$a_{n+1} - f_n a_n = g_n$$

$$\frac{a_{n+1}}{\prod_{k=0}^{n} f_k} - \frac{f_n a_n}{\prod_{k=0}^{n} f_k} = \frac{g_n}{\prod_{k=0}^{n} f_k}$$

$$\frac{a_{n+1}}{\prod_{k=0}^{n} f_k} - \frac{a_n}{\prod_{k=0}^{n-1} f_k} = \frac{g_n}{\prod_{k=0}^{n} f_k}$$

Let

$$A_n = \frac{a_n}{\prod_{k=0}^{n-1} f_k},$$

Then

$$A_{n+1} - A_n = \frac{g_n}{\prod_{k=0}^{n} f_k}$$

$$\sum_{m=0}^{n-1}(A_{m+1} - A_m) = A_n - A_0 = \sum_{m=0}^{n-1} \frac{g_m}{\prod_{k=0}^{m} f_k}$$

$$\frac{a_n}{\prod_{k=0}^{n-1} f_k} = A_0 + \sum_{m=0}^{n-1} \frac{g_m}{\prod_{k=0}^{m} f_k}$$

$$a_n = \left(\prod_{k=0}^{n-1} f_k\right)\left(A_0 + \sum_{m=0}^{n-1} \frac{g_m}{\prod_{k=0}^{m} f_k}\right)$$

If we apply the formula to $a_{n+1} = (1 + hf_{nh})a_n + hg_{nh}$ and take the limit h→o, we get the formula for first order linear differential equations with variable coefficients; the sum becomes an integral, and the product becomes the exponential function of an integral.

Solving general Homogeneous Linear Recurrence Relations

Many homogeneous linear recurrence relations may be solved by means of the generalized hypergeometric series. Special cases of these lead to recurrence relations for the orthogonal polynomials, and many special functions. For example, the solution to,

$$J_{n+1} = \text{—} J_n - J_{n-1}$$

is given by,

$$J_n = J_n(z),$$

the Bessel function, while

$$(b-n)M_{n-1} + (2n-b-z)M_n - nM_{n+1} = 0$$

is solved by,

$$M_n = M(n,b;z)$$

the confluent hypergeometric series.

Solving First-order Rational Difference Equations

A first order rational difference equation has the form $w_{t+1} = \frac{aw_t+b}{cw_t+d}$. Such an equation can be solved by writing w_t as a nonlinear transformation of another variable x_t which itself evolves linearly. Then standard methods can be used to solve the linear difference equation in x_t.

Stability

Stability of Linear Higher-order Recurrences

The linear recurrence of order *d*,

$$a_n = c_1 a_{n-1} + c_2 a_{n-2} + \cdots + c_d a_{n-d},$$

has the characteristic equation,

$$\lambda^d - c_1\lambda^{d-1} - c_2\lambda^{d-2} - \quad - c\ \lambda^0 = 0.$$

The recurrence is stable, meaning that the iterates converge asymptotically to a fixed value, if and only if the eigenvalues (i.e., the roots of the characteristic equation), whether real or complex, are all less than unity in absolute value.

Stability of Linear First-order Matrix Recurrences

In the first-order matrix difference equation,

$$[x_t - x^*] = A[x_{t-1} - x^*]$$

with state vector *x* and transition matrix *A*, *x* converges asymptotically to the steady state vector *x** if and only if all eigenvalues of the transition matrix *A* (whether real or complex) have an absolute value which is less than 1.

Stability of Nonlinear First-order Recurrences

Consider the nonlinear first-order recurrence,

$$x_n = f(x_{n-1}).$$

This recurrence is locally stable, meaning that it converges to a fixed point x^* from points sufficiently close to x^*, if the slope of f in the neighborhood of x^* is smaller than unity in absolute value: that is,

$$|f'(x^*)| < 1.$$

A nonlinear recurrence could have multiple fixed points, in which case some fixed points may be locally stable and others locally unstable; for continuous f two adjacent fixed points cannot both be locally stable.

A nonlinear recurrence relation could also have a cycle of period k for $k > 1$. Such a cycle is stable, meaning that it attracts a set of initial conditions of positive measure, if the composite function,

$$g(x) := f \circ f \circ \cdots \circ f(x)$$

with f appearing k times is locally stable according to the same criterion:

$$|g'(x^*)| < 1,$$

where x^* is any point on the cycle.

In a chaotic recurrence relation, the variable x stays in a bounded region but never converges to a fixed point or an attracting cycle; any fixed points or cycles of the equation are unstable.

Relationship to Differential Equations

When solving an ordinary differential equation numerically, one typically encounters a recurrence relation. For example, when solving the initial value problem,

$$y'(t) = f(t, y(t)), \ \ y(t_0) = y_0,$$

with Euler's method and a step size h, one calculates the values:

$$y_0 = y(t_0), \ \ y_1 = y(t_0 + h), \ \ y_2 = y(t_0 + 2h), \ldots$$

by the recurrence,

$$y_{n+1} = y_n + hf(t_n, y_n).$$

Systems of linear first order differential equations can be discretized exactly analytically.

Applications

Biology

Some of the best-known difference equations have their origins in the attempt to model population dynamics. For example, the Fibonacci numbers were once used as a model for the growth of a rabbit population.

The logistic map is used either directly to model population growth, or as a starting point for more detailed models of population dynamics. In this context, coupled difference equations are often used to model the interaction of two or more populations. For example, the Nicholson-Bailey model for a host-parasite interaction is given by:

$$N_{t+1} = \lambda N_t e^{-aP_t}$$
$$P_{t+1} = N_t(1-e^{-aP_t}),$$

with N_t representing the hosts, and P_t the parasites, at time t.

Integrodifference equations are a form of recurrence relation important to spatial ecology. These and other difference equations are particularly suited to modeling univoltine populations.

Computer Science

Recurrence relations are also of fundamental importance in analysis of algorithms. If an algorithm is designed so that it will break a problem into smaller subproblems (divide and conquer), its running time is described by a recurrence relation.

A simple example is the time an algorithm takes to find an element in an ordered vector with n elements, in the worst case.

A naive algorithm will search from left to right, one element at a time. The worst possible scenario is when the required element is the last, so the number of comparisons is n.

A better algorithm is called binary search. However, it requires a sorted vector. It will first check if the element is at the middle of the vector. If not, then it will check if the middle element is greater or lesser than the sought element. At this point, half of the vector can be discarded, and the algorithm can be run again on the other half. The number of comparisons will be given by:

$$c_1 = 1$$
$$c_n = 1 + c_{n/2}$$

which will be close to $\log_2(n)$.

Digital Signal Processing

In digital signal processing, recurrence relations can model feedback in a system, where outputs at one time become inputs for future time. They thus arise in infinite impulse response (IIR) digital filters.

For example, the equation for a "feedforward" IIR comb filter of delay T is:

$$y_t = (1-\alpha)x_t + \alpha y_{t-T}$$

Where x_t is the input at time t, y_t is the output at time t, and α controls how much of the delayed signal is fed back into the output. From this we can see that:

$$y_t = (1-\alpha)x_t + \alpha((1-\alpha)x_{t-T} + \alpha y_{t-2T})$$
$$y_t = (1-\alpha)x_t + (\alpha-\alpha^2)x_{t-T} + \alpha^2 y_{t-2T})$$

etc.

Economics

Recurrence relations, especially linear recurrence relations, are used extensively in both theoretical and empirical economics. In particular, in macroeconomics one might develop a model of various broad sectors of the economy (the financial sector, the goods sector, the labor market, etc.) in which some agents' actions depend on lagged variables. The model would then be solved for current values of key variables (interest rate, real GDP, etc.) in terms of exogenous variables and lagged endogenous variables.

This section contains the applications of formal power series to solving recurrence relations. Let us try to understand it using the following examples.

Example: 1. Determine a formula for the numbers a(n)'s, where a(n)'s satisfy the recurrence relation a(n) = 3 a(n – 1) + 2n, for n ≥ 1 with a(0) = 1.

Solution: Define $\mathrm{A}(x) = \sum_{n\geq 0} a(n)x^n$. Then using Example, one has:

$$\mathrm{A}(x) = \sum_{n\geq 0} a(n)x^n = a_0 + \sum_{n\geq 1} a(n)x^n = 1 + \sum_{n\geq 1}(3a(n-1)+2n)x^n.$$

$$= 3x\sum_{n\geq 1} a(n-1)x^{n-1} + 2\sum_{n\geq 1} nx^n + 1 = 3xA(x) + 2\frac{x}{(1-x)^2} + 1.$$

$$So,\ A(x) = \frac{1+x^2}{(1-3x)(1-x)^2} = \frac{5}{2(1-3x)} - \frac{1}{2(1-x)} - \frac{1}{(1-x)^2}.\ Thus,$$

$$a(n) = \left[x^n\right]A(x) = \frac{5}{2}3^n - \frac{1}{2} - (n+1) = \frac{5.3^n - 1}{2} - (n+1).$$

2. Determine a generating function for the numbers f (n) that satisfy the recurrence relation:

f (n) = f (n – 1) + f (n – 2), for n ≥ 2 with f (0) = 1 and f (1) = 1.

Hence or otherwise find a formula for the numbers f (n).

Solution: Define $F(x) = \sum_{n\geq 0} f(n)x^n$. Then one has

$$F(x)=\sum_{n\geq 0} f(n)x^n = 1+x\sum_{n\geq 2}(f(n-1)+f(n-2))x^n$$
$$=1+x+x\sum_{n\geq 2} f(n-1)x^{n-1}+x^2\sum_{n\geq 2} f(n-2)x^{n-2} = 1+xF(x)+x^2F(x).$$

Therefore, $F(x)=\frac{1}{1-x-x^2}$. Let $\alpha=\frac{1+\sqrt{5}}{2}$ and $\beta=\frac{1-\sqrt{5}}{2}$. Then it can be checked that $(1-\alpha x)(1-\beta x)=1-x-x^2$ and

$$F(x)=\frac{1}{\sqrt{5}}\left(\frac{\alpha}{1-\alpha x}-\frac{\beta}{1-\beta x}\right)=\frac{1}{\sqrt{5}}\left(\sum_{n\geq 0}\alpha^{n+1}x^n-\sum_{n\geq 0}\beta^{n+1}x^n\right).$$

Therefore,

$$f(n)=\left[x^n\right]F(x)=\frac{1}{\sqrt{5}}\sum_{n\geq 0}(\alpha^{n+1}-\beta^{n+1}).$$

As β < 0 and |β| < 1, we observe that $f(n)\approx\frac{1}{\sqrt{5}}\left(\frac{1+\sqrt{5}}{2}\right)^{n+1}$.

Remark: The numbers f (n), for n ≥ 0 are called Fibonacci numbers. It is related with the following problem: Suppose a couple bought a pair of rabbits (each one year old) in the year 2001. If a pair of rabbits starts giving birth to a pair of rabbits as soon as they grow 2 years old, determine the number of rabbits the couple will have in the year 2025.

3. Determine a formula for the numbers a(n)'s, where a(n)'s satisfy the recurrence relation a(n) = 3 a(n – 1) + 4a(n – 2), for n ≥ 2 with a(0) = 1 and a(1) = c, a constant.

Solution: Define $A(x)=\sum_{n\geq 0} a(n)x^n$. Then

$$A(x)=\sum_{n\geq 0} a(n)x^n = a_0+a_1x+\sum_{n\geq 2} a(n)x^n = 1+cx+\sum_{n\geq 2}(3a(n-1)+4a(n-2)x^n$$
$$=1+cx+3x\sum_{n\geq 2} a(n-1)x^{n-1}+4x^2\sum_{n\geq 2} a_{n-2}x^{n-2}$$
$$=1+cx+3x(A(x)-a_0)+4x^2A(x).$$

So, $A(x)=\frac{1+(c-3)x}{(1-3x-4x^2)}=\frac{1+(c-3)x}{(1-x)(1-4x)}$.

(a) If $c=4$ then $A(x)=\frac{1}{1-4x}$ and hence $a_n=\left[x^n\right]A(x)=4^n$..

(b) If $c\neq 4$ then $A(x)=\frac{1+c}{5}\cdot\frac{1}{1-4x}+\frac{4-c}{5}\cdot\frac{1}{1+x}$ and hence $a_n=\left[x^n\right]A(x)=\frac{(1+c)4^n}{5}+\frac{(-1)^n(4-c)}{5}$.

4. Determine a sequence, $\{a(n)\in\mathbb{R}:n\geq 0\}$, such that a_0 = 1 and $\sum_{k=0}^{n} a(k)a(n-k)=\binom{n+2}{2}$, for all n ≥ 1.

Solution: Define $A(x) = \sum_{n\geq 0} a(n)x^n$. Then, using the Cauchy product, one has:

$$A(x)^2 = \sum_{n\geq 0}\left(\sum_{k=0}^{n} a(k)a(n-k)\right)x^n = \sum_{n\geq 0}\binom{n+2}{2}x^n = \frac{1}{(1-x)^3}.$$

Hence, $A(x) = \frac{1}{(1-x)^{3/2}}$ and thus $a(n) = (-1)^n\binom{-3/2}{n} = \frac{3\cdot 5\cdot 7\cdots(2n+1)}{2^n n!}$, for all n ≥ 1.

5. Determine a generating function for the numbers $f(n,m), n,m \in \mathbb{Z}, n,m \geq 0$ that satisfy

$$f(n,m) = f(n-1,m) + f(n-1,m-1), (n,m) \neq (0,\ 0)$$
$$f(n,0) = 1, \textit{for all } n \geq 0 \textit{ and } f(0,m) = 0, \textit{for all } m > 0.$$

Hence or otherwise, find a formula for the numbers f (n, m).

Solution: Note that in the above recurrence relation, the value of m need not be ≤ n.

Method 1: Define $F_n(x) = \sum_{m\geq 0} f(n,m)x^m$. Then, for n ≥ 1, earlier equation gives:

$$\begin{aligned} F_n(x) &= \sum_{m\geq 0} f(n,m)x^m = \sum_{m\geq 0}\left(f(n-1,m) + f(n-1,m-1)\right)x^m \\ &= \sum_{m\geq 0} f(n-1,m)x^m + \sum_{m\geq 0} f(n-1,m-1)x^m \\ &= F_{n-1}(x) + xF_{n-1}(x) = (1+x)F_{n-1}(x) = \cdots = (1+x)^n F_0(x). \end{aligned}$$

Now, using the initial conditions, $F_0(x) = 1$ and hence $F_n(x) = (1+x)^n$. Thus,

$$f(n,m) = \left[x^m\right](1+x)^n = \binom{n}{m} \quad \textit{if } 0 \leq m \leq n \textit{ and} \qquad f(n,m) = 0, \textit{for} \qquad m > n.$$

Method 2: Define $G_m(y) = \sum_{n\geq 0} f(n,\ m)y^n$. *Then, for* $m \geq 1$, earlier equation gives:

$$\begin{aligned} G_m(y) &= \sum_{n\geq 0} f(n,m)y^n = \sum_{n\geq 0} (f(n-1,m) + f(n-1,m-1))y^n \\ &= \sum_{n\geq 0} f(n-1,\ m)y^n + \sum_{n\geq 0} f(n-1,\ m-1)y^n \\ &= yG_m(y) + yG_{m-1}(y). \end{aligned}$$

Therefore, $G_m(y) = \frac{1}{1-y}G_{m-1}(y)$. Now, using initial conditions, $G_0(y) = \frac{1}{1-y}$ and hence $G_m(y) = \frac{y^m}{(1-y)^{m+1}}$. Thus, $f(n.m) = \left[y^n\right]\frac{y^m}{(1-y)^{m+1}} = \left[y^{n-m}\right]\frac{1}{(1-y)^{m+1}} = \binom{n}{m}$, whenever 0 ≤ m ≤ n and f (n, m) = 0, for m > n.

Example: Determine a generating function for the numbers $S(n,m), n,m \in \mathbb{Z}, n,m \geq 0$ that satisfy

$$f(n,m) = f(n-1,m) + f(n-1,m-1), (n,m) \neq (0,\ 0)$$
$$f(n,0) = 1, \text{for all } n \geq 0 \text{ and } f(0,m) = 0, \text{ for all } m > 0.$$

Hence or otherwise find a formula for the numbers S(n, m).

Solution: Define $G_m(y) = \sum_{n\geq 0} S(n,m)y^n$. Then, for m ≥ 1, earlier equation gives:

$$\begin{aligned} G_m(y) &= \sum_{n\geq 0} S(n,m)y^n = \sum_{n\geq 0}(mS(n-1,m) + S(n-1,m-1))y^n \\ &= m\sum_{n\geq 0} S(n-1,m)y^n + \sum_{n\geq 0} S(n-1,m-1)y^n \\ &= myG_m(y) + yG_{m-1}(y). \end{aligned}$$

Therefore, $G_m(y) = \frac{y}{1-my}G_{m-1}(y)$. Using initial conditions, $G_0(y) = 1$ and hence:

$$G_m(y) = \frac{y^m}{(1-y)(1-2y)\cdots(1-my)} = y^m \sum_{k=1}^{m} \frac{\alpha_k}{1-ky},$$

where $\alpha_k = \frac{(-1)^{m-k}k^m}{k!(m-k)!}$, for 1 ≤ k ≤ m. Thus,

$$\begin{aligned} S(n,m) &= [y^n]\left(y^m \sum_{k=1}^{m} \frac{\alpha_k}{1-ky}\right) = \sum_{k=1}^{m} [y^{n-m}] \frac{\alpha_k}{1-ky} \\ &= \sum_{k=1}^{m} \alpha_k k^{n-m} = \sum_{k=1}^{m} \frac{(-1)^{m-k}k^n}{k!(m-k)!} \\ &= \frac{1}{m!}\sum_{k=1}^{m}(-1)^{m-k}k^n \binom{m}{k} = \frac{1}{m!}\sum_{k=1}^{m}(-1)^k(m-k)^n\binom{m}{k}. \end{aligned}$$

Therefore, $S(n,m) = \frac{1}{m!}\sum_{k=1}^{m}(-1)^k(m-k)^n\binom{m}{k}$ and $m!S(n,m) = \sum_{k=1}^{m}(-1)^k(m-k)^n\binom{m}{k}$.

This identity is generally known as the Stirling's Identity.

Observation:

(a) $H_n(x) = \sum_{m\geq 0} S(n,m)x^m$ is not considered. But verify that:

$$H_n(x) = (x + xD)^n \cdot 1 \text{ as } \quad H_0(x) = 1.$$

Therefore, $H_1(x) = x, H_2(x) = x + x^2, \cdots$. Hence, it is difficult to obtain a general formula for its coefficients. But it is helpful in showing that the numbers S(n, m),

for fixed n, first increase and then decrease (commonly called unimodal). The same holds for the sequence of binomial coefficients $\left\{\binom{n}{m}, m = 0,1,\ldots,n\right\}$.

(b) Since there is no restriction on the non-negative integers n and m, the expression Equation (3) is also valid for n < m. But, in this case, we know that S(n, m) = 0. Hence, verify that $\sum_{k=1}^{m}\frac{(-1)^{m-k}k^{n-1}}{(k-1)!(m-k)!}=0$, whenever n < m.

2. Bell Numbers: For a positive integer n, the nth Bell number, denoted b(n), is the number of partitions of the set {1, 2, . . . , n}. Therefore, by definition, $b(n)=\sum_{m=1}^{m}S(n,m)$,for n ≥ 1 and by convention, b(0) = 1. Thus, for n ≥ 1,

$$b(n)=\sum_{m=1}^{m}S(n,m)=\sum_{m\geq 1}S(n,m)=\sum_{m\geq 1}\sum_{k=1}^{m}\frac{(-1)^{m-k}k^{n-1}}{(k-1)!(m-k)!}$$
$$=\sum_{k\geq 1}\frac{k^n}{k!}\sum_{m\geq k}\frac{(1-)^{m-k}}{(m-k)!}=\frac{1}{e}\sum_{k\geq 1}\frac{k^n}{k!}.$$

Note that the above equation is valid even for n = 0. Also, observe that b(n) has terms of the form $\frac{k^n}{k!}$ and hence we compute its exponential generating function.

Thus, if $B(x)=\sum_{n\geq 0}b(n)\frac{x^n}{n!}$ then

$$B(x)=1+\sum_{n\geq 1}b(n)\frac{x^n}{n!}=1+\sum_{n\geq 1}\left(\frac{1}{e}\sum_{k\geq 1}\frac{k^n}{k!}\right)\frac{x^n}{n!}$$
$$=1+\frac{1}{e}\sum_{k\geq 1}\frac{1}{k!}\sum_{n\geq 1}k^n\frac{x^n}{n!}=1+\frac{1}{e}\sum_{k\geq 1}\frac{1}{k!}\sum_{n\geq 1}\frac{(kx)^n}{n!}$$
$$=1+\frac{1}{e}\sum_{k\geq 1}\frac{1}{k!}(e^{kx}-1)=1+\frac{1}{e}\sum_{k\geq 1}\left(\frac{(e^x)^k}{k!}-\frac{1}{k!}\right)$$
$$=1+\frac{1}{e}(e^{e^x}-1-(e-1))=e^{e^x}-1.$$

Recall that $e^{e^x}-1$ is a valid formal power series. Now, let us derive the recurrence relation for b(n)'s. Taking the natural logarithm on both the sides of above equation, one has $Ln\left(\sum_{n\geq 0}b(n)\frac{x^n}{n!}\right)=e^x-1$.

Now, differentiation with respect to x gives $\frac{1}{\sum_{n\geq 0}b(n)\frac{x^n}{n!}}.\sum_{n\geq 0}b(n)\frac{x^{n-1}}{(n-1)!}=e^x$. Therefore, after cross multiplication and a multiplica-tion with x, implies:

$$\sum_{n\geq 1}\frac{b(n)x^n}{(n-1)!}=xe^x\sum_{n\geq 0}b(n)\frac{x^n}{n!}=x\left(\sum_{m\geq 0}\frac{x^m}{m!}\right)\cdot\left(\sum_{n\geq 0}b(n)\frac{x^n}{n!}\right).$$

Thus,

$$\frac{b(n)}{(n-1)!}=\left[x^n\right]\frac{b(n)x^n}{(n-1)!}=\left[x^n\right]x\left(\sum_{m\geq 0}\frac{x^m}{m!}\right)\cdot\left(\sum_{n\geq 0}b(n)\frac{x^n}{n!}\right)=\sum_{m=0}^{n-1}\frac{1}{(n-1-m)!}\cdot\frac{b(m)}{m!}$$

Hence, it follows that $b(n)=\sum_{m=0}^{n-1}\binom{n-1}{m}b(m)$, for n ≥ 1, with b(0) = 1.

Example: Determine the number of ways of arranging n pairs of parentheses (left and right) such that at any stage the number of right parentheses is always less than or equal to the number of left parentheses.

Solution: Recall that this number equals C_n, the nth Catalan number. Let us obtain a recurrence relation for these numbers and use it to get a formula for C_n's.

Let P_n denote the arrangements of those n pairs of parentheses that satisfy "at any stage, the number of left parentheses is always greater than or equal to the number of right parentheses". Then $|P_n| = C_n$, for all n ≥ 1. Also, let Q_n denote those elements of P_n for which, "at the 2k-th stage, for k < n, the number of left parentheses is strictly greater than the number of right parentheses".

We now claim that $|Q_1| = 1$ and $|Q_n| = |P_{n-1}|$, for n ≥ 2.

Clearly $|Q_1| = 1$. Note that, for n ≥ 2, any element of Q_n, necessarily starts with two left parentheses and ends with two right parentheses. So, if we remove the first left parenthesis and the last right parenthesis from each element of Q_n then one obtains an element of P_{n-1}. In a similar way, if we add one left parenthesis at the beginning and a right parenthesis at the end of an element of P_{n-1}, one obtains an element of Q_n. Hence, there is one-to-one correspondence between the set Q_n and P_{n-1}. Thus, $|Q_n| = |P_{n-1}| = C_{n-1}$.

Let n ≥ 2 and consider an element of P_n. Then, for some k, 1 ≤ k ≤ n, the first k pairs of parentheses will have the property that the number of left parentheses is strictly greater than the number of right parentheses, for 1 ≤ ℓ < k, i.e., they form an element of Q_k and the remaining (n – k) pairs of parentheses will form an element of P_{n-k}. Hence, if $|P_0| = |Q_1| = 1$, one has $C_n=|P_n|=\sum_{k=1}^{n}|Q_k||P_{n-k}|=\sum_{k=1}^{n}|P_{k-1}||P_{n-k}|=\sum_{k=1}^{n}C_{k-1}C_{n-k},\ for\ n\geq 2$. Now, define $C(x)=\sum_{n\geq 0}C_nx^n$. Then

$$\begin{aligned}C(x)&=\sum_{n\geq 0}C_nx^n=1+\sum_{n\geq 1}C_nx^n=1+\sum_{n\geq 1}\left(\sum_{k=1}^{n}C_{k-1}C_{n-k}\right)x^n\\&=1+x\left(\sum_{k\geq 1}C_{k-1}x^{k-1}\sum_{n\geq k}C_{n-k}x^{n-k}\right)=1+x\left(C(x)\sum_{k\geq 1}C_{k-1}x^{k-1}\right)\\&=1+x(C(x))^2.\end{aligned}$$

Thus, $xC(x)^2-C(x)+1=0$ and $C(x)=\frac{1\pm\sqrt{1-4x}}{2x}$. Therefore, using $C_0=1$, one obtains $C(x)=\frac{1-\sqrt{1-4x}}{2x}$. Hence,

$$C_n = \left[x^n\right]C(x) = \frac{1}{2}\cdot\left[x^{n+1}\right](1-\sqrt{1-4x})$$

$$= -\frac{1}{2}\cdot\frac{\frac{1}{2}\left(\frac{1}{2}-1\right)\left(\frac{1}{2}-2\right)\cdots\left(\frac{1}{2}-n\right)}{(n+1)!}(-4)^{n+1}$$

$$= 2(-4)^{\Omega}\cdot\frac{1\cdot(-1)\cdot(-3)\cdot(-5)\cdots(1-2n)}{2^{n+1}(n+1)!} = 2^n\frac{1\cdot 3\cdot 5\cdot\cdots\cdot(2n-1)}{(n+1)!}$$

$$= \frac{1}{n+1}\binom{2n}{n}, \textit{ the } n^{th} \textit{ Catalan Number.}$$

The ideas learnt in the previous sections will be used to get closed form expressions for sums arising out of binomial coefficients.

Example: Find a closed form expression for the numbers $a(n) = \sum_{k\geq 0}\binom{k}{n-k}$.

Solution: Define $A(x) = \sum_{n\geq 0} a(n)x^n$. Then

$$A(x) = \sum_{n\geq 0} a(n)x^n = \sum_{n\geq 0}\left(\sum_{k\geq 0}\binom{k}{n-k}\right)x^n = \sum_{k\geq 0}\left(\sum_{n\geq 0}\binom{k}{n-k}x^n\right)$$

$$= \sum_{k\geq 0} x^k\left(\sum_{n\geq k}\binom{k}{n-k}x^{n-k}\right) = \sum_{k\geq 0} x^k(1+x)^k = \sum_{k\geq 0} x^k(x(1+x))^k = \frac{1}{1-x(1-x)}.$$

Therefore, Example implies $a(n) = \left[x^n\right]A(x) = \left[x^n\right]\frac{1}{1-x(1+x)} = F_n$, the n-th Fibonacci number.

2. Find a closed form expression for the polynomials $a(n,x) = \sum_{k=0}^{\lfloor \frac{n}{2} \rfloor}\binom{n-k}{k}(-1)^k x^{n-2k}$.

Solution: Define $A(x,y) = \sum_{n\geq 0} a(n,x)y^n$. Then

$$A(x,y) = \sum_{n\geq 0} a(n,x)y^n = \sum_{n\geq 0}\left(\sum_{k=0}^{\lfloor \frac{n}{2} \rfloor}\binom{n-k}{k}(-1)^k x^{n-2k}\right)y^n$$

$$= \sum_{k\geq 0}(-1)^k y^{2k}\left(\sum_{n\geq 2k}\binom{n-k}{k}(xy)^{n-2k}\right)$$

$$= \sum_{k\geq 0}(-1)^k y^{2k}(xy)^{-k}\left(\sum_{t\geq k}\binom{t}{k}(xy)^t\right)$$

$$= \sum_{k\geq 0}(-y^2)^k(xy)^{-k}\frac{(xy)^k}{(1-xy)^{k+1}} = \frac{1}{1-xy}\cdot\sum_{k\geq 0}\left(\frac{(-y^2)}{1-xy}\right)^k$$

$$= \frac{1}{1-xy}\cdot\frac{1}{1-\frac{-y^2}{1-xy}} = \frac{1}{1-xy+y^2} = \frac{1}{(1-\alpha y)(1-\beta y)},$$

where $\alpha = \frac{x+\sqrt{x^2-4}}{2}$ and $\beta = \frac{x-\sqrt{x^2-4}}{2}$. Thus,

$$a(n,x)=[y^n]A(x,y)=[y^n]\frac{1}{1-xy+y^2}=[y^n]\frac{1}{\alpha-\beta}\left(\frac{\alpha}{1-\alpha y}-\frac{\beta}{1-\beta y}\right)$$
$$=\frac{1}{\alpha-\beta}\left(\alpha^{n+1}-\beta^{n+1}\right)$$
$$=\frac{1}{\sqrt{x^2-4}}\left(\left(\frac{x+\sqrt{x^2-4}}{2}\right)^{n+1}-\left(\frac{x-\sqrt{x^2-4}}{2}\right)^{n+1}\right).$$

Since α and β are the roots of $y^2-xy+1=0$, $\alpha^2=\alpha x-1$ and $\beta^2=\beta x-1$. Therefore, verify that the $a(n,x)$'s satisfy the recurrence relation $a(n,x)=xa(n-1,x)-a(n-2,x),\ for\ n\geq 2,$ with initial conditions $a(0,x)=1$ and $a(1,x)=x$.

Let A = (a_{ij}) be an n × n matrix, with a_{ij} = 1, whenever |i− j| = 1 and 0, otherwise. Then A is an adjacency matrix of a tree T on n vertices, say 1, 2, . . . , n with the vertex i being adjacent to i + 1, for 1 ≤ i ≤ n − 1. It can be verified that if $a(n,x)=det(xI_n-A)$, the characteristic polynomial of A, then $a(n,x)$'s satisfy the above recurrence relation. The polynomials $a(n,2x)$'s are also known as Chebyshev's polynomial of second kind.

Let us now substitute different values for x to obtain different expressions and then use them to get binomial identities.

1. Let $x=z+\frac{1}{z}$. Then $\sqrt{x^2-4}=z-\frac{1}{z}$ and we obtain $a\left(n,z+\frac{1}{z}\right)=\frac{z^{2n+2}-1}{(z^2-1)z^n}$. Hence, we have

$\sum_{k=0}^{\lfloor\frac{n}{2}\rfloor}\binom{n-k}{k}(-1)^k\left(z+\frac{1}{z}\right)^{n-2k}=\frac{z^{2n+2}-1}{(z^2-1)z^n}$. Or equivalently,

$$\sum_{k=0}^{\lfloor\frac{n}{2}\rfloor}\binom{n-k}{k}(-1)^k(z^2+1)^{n-2k}z^{2k}=\frac{z^{2n+2}-1}{z^2-1}.$$

2. Writing x in place of z^2, we obtain the following identity.

$$\sum_{k=0}^{\lfloor\frac{n}{2}\rfloor}\binom{n-k}{k}(-1)^k(x+1)^{n-2k}x^k=\frac{x^{n+1}+1}{x-1}=\sum_{k=0}^{n}x^k.$$

3. Hence, equating the coefficient of x^m in the above equation gives the identity

$$\sum_{k=0}^{\lfloor\frac{n}{2}\rfloor}(-1)^k\binom{n-k}{k}\binom{n-2k}{m-k}=\begin{cases}1, & if\ \ 0\leq m\leq n;\\ 0, & otherwise.\end{cases}$$

4. Substituting $x=1$ gives $\sum_{k=0}^{\lfloor \frac{n}{2} \rfloor}(-1)^k\binom{n-k}{k}2^{n-2k}=n+1.$

Example: Determine the generating function for the numbers $a(n,y)=\sum_{k\geq 0}\binom{n+k}{2k}y^k$.

Solution: Define $A(x,y)=\sum_{n\geq 0}a(n,y)x^n$. Then

$$A(x,y)=\sum_{n\geq 0}\left(\sum_{k\geq 0}\binom{n+k}{2k}y^k\right)x^n=\sum_{k\geq 0}\left(\frac{y}{x}\right)^k\sum_{n\geq k}\binom{n+k}{2k}x^{n+k}$$

$$=\sum_{k\geq 0}\left(\frac{y}{x}\right)^k\frac{x^{2k}}{(1-x)^{2k+1}}=\frac{1}{1-x}\sum_{k\geq 0}\left(\frac{yx}{(1-x)^2}\right)^k$$

$$=\frac{1-x}{(1-x)^2-xy}.$$

1. Verify that if we substituting y = −2 then

$$\sum_{k\geq 0}\binom{n+k}{2k}(-2)^k=\left[x^n\right]A(x,-2)=\left[x^n\right]\frac{1-x}{(1+x^2)}=(-1)^{\lceil n/2\rceil}.$$

2. Verify that if we substituting y = −4 then

$$\sum_{k\geq 0}\binom{n+k}{2k}(-4)^k=\left[x^n\right]A(x,-4)=\left[x^n\right]\frac{1-x}{(1+x)^2}=(-1)^n(2n+1).$$

3. Let $f(n)=\sum_{k\geq 0}\binom{n+k}{2k}2^{n-k}$ and let $F(z)=\sum_{n\geq 0}f(n)z^n$. Then verify that

$$F(z)=A\left(2z,\frac{1}{2}\right)=\frac{1-2z}{(1-z)(1-4z)}=\frac{2}{3}\cdot\frac{1}{1-4z}+\frac{1}{3}\cdot\frac{1}{1-z}.$$

Hence, $f(n)=\left[z^n\right]F(z)=\frac{2.4^n}{3}+\frac{1}{3}=\frac{2^{2n+1}+1}{3}.$

References

- Wang, Xiang-Sheng; Wong, Roderick (2012). "Asymptotics of orthogonal polynomials via recurrence relations". Anal. Appl. 10 (2): 215–235. doi:10.1142/S0219530512500108
- Björklund, A.; Husfeldt, T.; Koivisto, M. (2009), "Set partitioning via inclusion–exclusion", SIAM Journal on Computing, 39 (2): 546–563, doi:10.1137/070683933
- Berstel, Jean; Reutenauer, Christophe (2011). Noncommutative rational series with applications. Encyclopedia

of Mathematics and Its Applications. 137. Cambridge: Cambridge University Press. ISBN 978-0-521-19022-0. Zbl 1250.68007

- Jacques, Ian (2006). Mathematics for Economics and Business (Fifth ed.). Prentice Hall. pp. 551–568. ISBN 0-273-70195-9
- Thomas H. Cormen, Charles E. Leiserson, Ronald L. Rivest, and Clifford Stein. Introduction to Algorithms, Second Edition. MIT Press and McGraw-Hill, 1990. ISBN 0-262-03293-7

Permissions

All chapters in this book are published with permission under the Creative Commons Attribution Share Alike License or equivalent. Every chapter published in this book has been scrutinized by our experts. Their significance has been extensively debated. The topics covered herein carry significant information for a comprehensive understanding. They may even be implemented as practical applications or may be referred to as a beginning point for further studies.

We would like to thank the editorial team for lending their expertise to make the book truly unique. They have played a crucial role in the development of this book. Without their invaluable contributions this book wouldn't have been possible. They have made vital efforts to compile up to date information on the varied aspects of this subject to make this book a valuable addition to the collection of many professionals and students.

This book was conceptualized with the vision of imparting up-to-date and integrated information in this field. To ensure the same, a matchless editorial board was set up. Every individual on the board went through rigorous rounds of assessment to prove their worth. After which they invested a large part of their time researching and compiling the most relevant data for our readers.

The editorial board has been involved in producing this book since its inception. They have spent rigorous hours researching and exploring the diverse topics which have resulted in the successful publishing of this book. They have passed on their knowledge of decades through this book. To expedite this challenging task, the publisher supported the team at every step. A small team of assistant editors was also appointed to further simplify the editing procedure and attain best results for the readers.

Apart from the editorial board, the designing team has also invested a significant amount of their time in understanding the subject and creating the most relevant covers. They scrutinized every image to scout for the most suitable representation of the subject and create an appropriate cover for the book.

The publishing team has been an ardent support to the editorial, designing and production team. Their endless efforts to recruit the best for this project, has resulted in the accomplishment of this book. They are a veteran in the field of academics and their pool of knowledge is as vast as their experience in printing. Their expertise and guidance has proved useful at every step. Their uncompromising quality standards have made this book an exceptional effort. Their encouragement from time to time has been an inspiration for everyone.

The publisher and the editorial board hope that this book will prove to be a valuable piece of knowledge for students, practitioners and scholars across the globe.

Index

www.ingramcontent.com/pod-product-compliance
Lightning Source LLC
LaVergne TN
LVHW080724210926
848485LV00003B/517